Governing Sustainable Seafood

Taking a social science approach, this book explores the governance of sustainable seafood, which is fundamental to food and nutrition security as well as being an important source of income and employment in many regions.

Due to the importance of protein and other fishery and aquaculture by-products, many wild fisheries are coming under pressure, and this increasing demand has created a strong driver to expand aquaculture. As a result, the social and environmental sustainability of these production systems have come into question. The authors of the book explore the governance of sustainable seafood, taking into account the rise of social movements through environmental non-governmental organisations, the nature and perceived limits of government regulation within and beyond the state, and the promise of market-based approaches to governance such as ecolabelling. The book focuses on how concern over sustainable seafood has been translated into different current forms of governance. It then assesses what alternative governance approaches are starting to emerge that combine movements, states and markets for sustainable seafood production and consumption, and their effects. The book concludes with a vision for the future through key principles for evaluating the collective impact of governing sustainable seafood.

This timely volume will be key reading for researchers interested in fisheries and aquaculture governance, as well as coastal and marine policies and sustainable food movements more broadly. It will also be of interest to practitioners and policymakers engaged in creating fishery policies and sustainable fishery development.

Simon R. Bush is professor and chair of the Environmental Policy Group at Wageningen University, the Netherlands. He is also associate editor of the journals *Maritime Studies* and *Environmental Development.*

Peter Oosterveer is professor of global food governance in the Environmental Policy Group at Wageningen University, the Netherlands. He is the author of *Food, Globalization and Sustainability* (Routledge, 2011).

Other books in the Earthscan Food and Agriculture Series

The New Peasantries
Rural Development in Times of Globalization (Second Edition)
Jan Douwe van der Ploeg

Sustainable Intensification of Agriculture
Greening the World's Food Economy
Jules N. Pretty and Zareen Pervez Bharucha

Contested Sustainability Discourses in the Agrifood System
Edited by Douglas H. Constance, Jason Konefal and Maki Hatanaka

The Financialization of Agri-Food Systems
Contested Transformations
Edited by Hilde Bjørkhaug, André Magnan and Geoffrey Lawrence

Redesigning the Global Seed Commons
Law and Policy for Agrobiodiversity and Food Security
Christine Frison

Organic Food and Farming in China
Top-down and Bottom-up Ecological Initiatives
Steffanie Scott, Zhenzhong Si, Theresa Schumilas and Aijuan Chen

Farming, Food and Nature
A Sustainable Future for Animals, People and the Environment
Edited by Joyce D'Silva and Carol McKenna

Governing Sustainable Seafood
Simon R. Bush and Peter Oosterveer

For further details please visit the series page on the Routledge website: www.routledge.com/books/series/ECEFA/

Governing Sustainable Seafood

Simon R. Bush and Peter Oosterveer

Routledge
Taylor & Francis Group
LONDON AND NEW YORK

earthscan
from Routledge

First published 2019 by Routledge

2 Park Square, Milton Park, Abingdon, Oxfordshire OX14 4RN
52 Vanderbilt Avenue, New York, NY 10017

Routledge is an imprint of the Taylor & Francis Group, an informa business

First issued in paperback 2020

British Library Cataloguing-in-Publication Data
A catalogue record for this book is available from the British Library

Library of Congress Cataloging-in-Publication Data
Names: Oosterveer, Peter, author.
Title: Governing sustainable seafood / Peter Oosterveer and Simon Bush.
Description: Milton Park, Abingdon, Oxon ; New York, NY : Routledge, 2019. | Series: Earthscan food and agriculture | Includes bibliographical references and index.
Identifiers: LCCN 2018045807| ISBN 9781138017542 (hbk) | ISBN 9781315780429 (ebk)
Subjects: LCSH: Sustainable aquaculture–Social aspects. | Sustainable aquaculture–Environmental aspects. | Sustainable fisheries–Social aspects. | Seafood industry–Social aspects.
Classification: LCC SH136.S88 O57 2019 | DDC 639.8028/6–dc23
LC record available at https://lccn.loc.gov/2018045807

ISBN: 978-1-138-01754-2 (hbk)
ISBN: 978-0-367-62854-3 (pbk)

Typeset in Bembo
by Wearset Ltd, Boldon, Tyne and Wear

Contents

Illustrations

Figures

Table

Boxes

Preface

Sustainable seafood is central to most of our work, but finishing this book has proven more challenging than we could have ever imagined at the start of this project four years ago.

Our interest in the sustainable seafood movement directly reflects the number of standards, codes and initiatives it entails. Each new method or approach taken up involves fishers, non-governmental organisations (NGOs), governments and retailers in novel combinations and with differing goals and aspirations. The challenge for two academics following these developments is the pace at which the sustainable movement continues to develop. It takes serious effort to keep abreast of all the changes and innovation, let alone document and make sense of what these changes mean for the ultimate goal of delivering sustainable seafood outcomes.

The challenge of covering the full range of sustainability initiatives is therefore incomplete. Our goal in this book has instead been to provide enough detail for both academics and practitioners to reflect on the social complexity of governing global food systems. This, we hope, is the enduring value of the theoretical project this book entails. We hope, however, that this 'theory' will not scare away those with a more central interest in seafood, fisheries and aquaculture. Instead, our goal is to inspire readers from all disciplines (and even practitioners) to reimagine and more effectively design their goals for sustainable seafood provision to achieve greater impact.

As the chapters unfold it will become apparent how much we have drawn on our earlier work and the work of our close co-authors and students over the last decade. We are indebted to all of these colleagues, both within and across the social and natural sciences, for helping us make sense of what shapes global fisheries and aquaculture production, trade and consumption. We hope we have represented your ideas accurately, and challenge you to think further with us on the future of food system governance – whether it be for fish or any other global food commodity.

Simon Bush and Peter Oosterveer, Wageningen, September 2018

Abbreviations

AIP	aquaculture improvement project
ASC	Aquaculture Stewardship Council
BAP	Best Aquaculture Practices
CBL	Dutch Retail Association
CITES	Convention on International Trade in Endangered Species
CSF	community-supported fishery
EEZ	exclusive economic zone
EII	Earth Island Institute
EJF	Environmental Justice Foundation
EMPA	Entrepreneurial Marine Protected Areas
ESA	Endangered Species Act (USA)
EU	European Union
FAO	Food and Agriculture Organization of the United Nations
FIP	fishery improvement project
FoS	Friends of the Sea
GAA	Global Aquaculture Alliance
GAP	Good Aquaculture Practices
GATT	General Agreement on Trade and Tariffs
GCC	global commodity chain
GDP	gross domestic product
GGN	GlobalGAP number
GIZ	German International Development Organisation
GPN	global production network
GPS	global positioning system
GSI	Global Salmon Initiative
GSSI	Global Sustainable Seafood Initiative
GVC	global value chain
HACCP	hazard analysis and critical control points
HCR	harvest control rules
ICT	information and communications technology
IDH	Netherlands Sustainable Trade Initiative
ISO	International Organization for Standardization
ISSF	International Seafood Sustainability Foundation

ITQ	individual transferable quota
IUU	illegal, unreported and unregulated (fisheries/fishing)
MSC	Marine Stewardship Council
MT	metric tonne
NGO	non-governmental organisation
OECD	Organisation for Economic Co-operation and Development
RFM	Responsible Fisheries' Management
RFMO	regional fisheries' management organisation
SDG	Sustainable Development Goal
SFP	Sustainable Fisheries Partnership
SPS	sanitary and phytosanitary measures
SSA	sustainable seafood aggregator
SSC	Sustainable Seafood Coalition
TAC	total allowable catch
TBT	Technical Barriers to Trade
UN	United Nations
UNCLOS	UN Convention on the Law of the Seas
US	United States
WTO	World Trade Organization
WWF	World Wide Fund for Nature

1 Governing seafood

An introduction

'Seafood' refers to a wide range of finfish, shellfish and crustacean species harvested and farmed in both marine and freshwater environments. Unlike other food sectors, seafood is divided between the wild harvest and the farming of hundreds of different species of fish and other aquatic species by both industrial and artisanal fishers and farmers in all countries of the world. The scope of fishery production is impressive, ranging from tuna fisheries around the world's oceans to anchovy fisheries off the coast of Peru, white fish stocks in the North Atlantic, inland fisheries of the East African lakes and reef and demersal fisheries throughout the tropics. The bulk of aquaculture production occurs in the tropics, especially in Asia. However, as this sector has grown so too has its geographical area, with production now extending to the colder climates of North and South America and Europe.

The trade and consumption of seafood are also global, with approximately half of the seafood consumed by developed country markets, like the United States (US) and the European Union (EU), originating in developing countries (Pramod et al., 2014; EUMOFA, 2016). At the same time, domestic trade and consumption in developing countries make a fundamental contribution to the nutrition and food security of vast swathes of rural and urban populations. Estimates by Belton et al. (2017) show that this 'invisible' domestic consumption could account for as much as 90 per cent of total seafood production in developing countries.

It is the global nature of seafood trade, however, that has brought the sustainability of fisheries and aquaculture into the spotlight. Based on their experience with food safety, regulators and consumers now expect to know where, how and by whom seafood is produced. In the last decade of the twentieth century, demands for such knowledge shifted the attention of these markets to questions of environmental sustainability, driven by public debates in the US and EU around a number of high-profile cases of unsustainable seafood production, including the collapse of cod stocks in the North Atlantic, the widely reported maiming and death of thousands of dolphins by tuna purse seiners in the Eastern Pacific, revelations of widespread illegal fishing for Chilean seabass in the South Pacific and the impacts of shrimp and salmon aquaculture on coastal habitats and biodiversity.

NGO campaigns have been successful in raising public concern over the impact seafood consumption practices have on fish stocks and coastal environments. They have also demonstrated the complexity and opaque nature of international seafood production and trade. As NGOs dug deeper it became apparent that the capacity of exporting states to produce their seafood supplies sustainably was weak (Smith et al., 2010). After nearly three decades, these various civil society concerns have culminated in what is today referred to as the 'sustainable seafood movement'.

The sustainable seafood movement has been shaped by partnerships between NGOs, US-based family philanthropic foundations and industry, all of which were in search of solutions to the perceived failures of governments to manage the sector (Safina, 1995). Their combined efforts have focused on the development of norms, principles and standards that attempt to define and codify sustainability in the form of third-party certification, like the Marine Stewardship Council (MSC), and NGO recommendation lists like Monterey Bay Aquarium's Seafood Watch. These norms, principles and standards have been translated into evidence-based criteria aimed overwhelmingly at assessing the process of production (Hatanaka et al., 2005). As outlined by Roheim et al. (2018), the prevailing 'theory of change' behind these initiatives is to use the market to incentivise demonstrated compliance with these assessment criteria, through either a price premium or preferential access to major (export) markets.

The initiatives assessing sustainability in the seafood sector now include around 50 standards, codes, recommendation lists and traceability programmes (Jacquet and Pauly, 2007; Parkes et al., 2009; Roheim, 2009). There are three main reasons for this proliferation. First, sustainability is an imperfectly defined product attribute, allowing new 'niches' to emerge as sustainability is redefined and expanded to new issues and concerns. Second, new initiatives have been designed to reach a critical mass of producers – especially those in developing countries which have proven more difficult to enrol in the seafood movement. Third is the emergence of meta-governance arrangements or 'standards of standards' that have provided a framework for the development and professionalisation of sustainability initiatives.

Central to the success of these initiatives is their ability to establish and maintain legitimacy to set and enforce norms, standards and principles over seafood production. However, the legitimacy of these private standards is increasingly contested. If their methods for setting or assessing sustainability are not shown to be adequate, or they are not able to engage a critical mass of producers, or they are unable to demonstrate positive impacts on sustainability, their societal credibility is diminished (Bush et al., 2013). Being able to address all three of these dimensions goes far beyond the technical dimensions of standards and assessment methodologies. It brings us to the question of governance – that is, whether and how these initiatives, either individually or in combination, set the social and technical conditions for actors in the seafood industry to deliberate and act towards solving the societal challenge of sustainability.

It is precisely this question of governance that we address in this book. We aim to identify the main challenges faced by the global seafood movement, and to develop a new theoretical approach to understand better how the seafood movement can shape production, trade and consumption moving forward. By addressing these goals, we attempt to shift the prevailing focus of academics and practitioners away from the technical dimensions of codes, standards, surveillance and assessment and instead to understand how a *social* seafood movement enables the kind of creative innovation and learning needed to contribute to the sustainable provision of seafood under conditions of globalisation.

In the remainder of this chapter we further describe the importance of the sustainability challenges faced by the seafood industry. We also set out the case for an understanding of the seafood movement as a global governance problem, before outlining the structure of the subsequent chapters.

Global seafood trends

The seafood industry is made up of two distinct sectors: capture fisheries and aquaculture. Capture fisheries involve the use of different methods and materials (equipment and boats) in an area of water or seabed to harvest fish and other aquatic organisms (Fletcher et al., 2002). Aquaculture, in contrast, involves the farming of aquatic organisms based on some form of intervention in the rearing process to enhance production, such as regular stocking, feeding and protection from predators (Edwards and Demaine, 1998). Taken together, fisheries and aquaculture have contributed to the impressive growth of seafood production over the last 60 years, rising from 19 million metric tonnes (MT) in 1950 to 171 million MT in 2016 (FAO, 2018a) (Figure 1.1). However, each sector has a very different growth trajectory when looked at separately.

Capture fisheries' production

According to the Food and Agriculture Organization of the United Nations (FAO) (ibid.), global capture fisheries' production peaked in 1996 at around 96 million MT. By 2016 total marine catch was 79.3 million MT, down almost two million MT from 81.2 million MT in 2015. Inland fisheries provided an additional 11.6 million MT in 2016, representing 12.8 per cent of total catches (ibid.).

The accuracy of these figures is disputed, however. Catch reconstructions by the Sea Around Us project at the University of British Columbia indicate that catches between 1950 and 2010 were 50 per cent higher than reported by the FAO (ibid.), peaking at 130 million MT in the mid-1990s and declining at a rate three times faster than reported by the FAO ever since. These reconstructions also show differences between fleets. While industrial catch

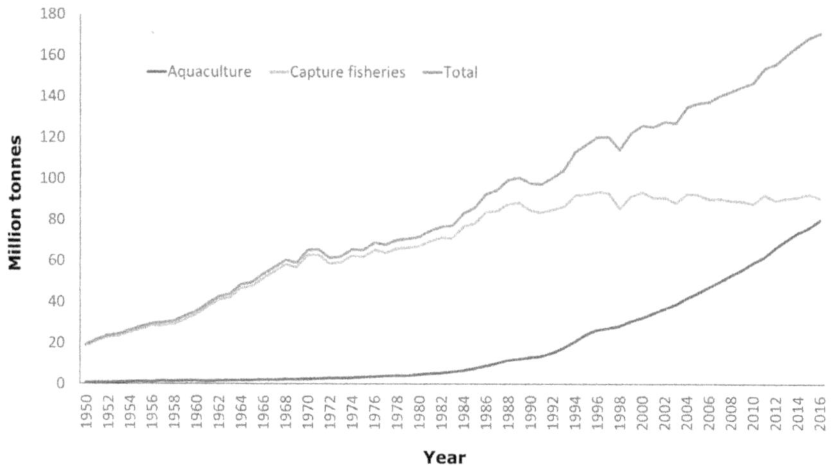

Figure 1.1 Global seafood production, 1950–2016.

Source: based on data from FAO (2018a).

Note
Figures are for animal protein, minus plants and marine mammal production.

appears to have declined, catches from small-scale coastal fisheries demonstrate slight growth.

The exact status of global fish stocks is also contested. The most recent assessment from the FAO (ibid.) shows that global landings of wild-caught fish are thought to be close to, or even exceed, their estimated sustainability levels. These official statistics estimate that 'the fraction of marine fish stocks fished within biologically sustainable levels has exhibited a decreasing trend, from 90.0 percent in 1974 to 66.9 percent in 2015'. At the same time, the proportion of

> stocks fished at biologically unsustainable levels increased from 10 percent in 1974 to 33.1 percent in 2015 … [when in the same year] maximally sustainably fished stocks (formerly termed fully fished stocks) accounted for 59.9 percent and under-fished stocks for 7.0 percent of the total assessed stocks.
>
> (Ibid.: 40)

The forecast for increasing the number of stocks fished at sustainable levels also remains unclear. Some assessments show that although two-thirds of stocks have a biomass below what would produce maximum sustainable yields, nearly half of these fisheries have taken measures to lower exploitation rates to levels that would support recovery (Worm et al., 2009; Costello et al., 2012). Countering this, the FAO (2018b) reports that any such trend for recovering fish stocks that are currently overfished is questionable, given that two to three times of a species lifespan is needed to bring it back to healthy levels of abundance.

Aquaculture production

In contrast to capture fisheries, aquaculture has continuously expanded over the last five decades to reach 80 million MT of food fish, 30.1 million MT of aquatic plants and 38,000 MT of non-food products by 2016 (FAO, 2018b). Two-thirds of this overall production are finfish, and the rest is made up of a variety of molluscs, crustaceans and other aquatic animals. A further division can be made between extractive and non-extractive species. Marine bivalves and seaweeds are extractive, in that they remove waste materials and nutrient loads from the water column, including waste from fed species. These extractive species account for half of total global production, but play a far smaller role in seafood trade and consumption than finfish.

It is not only the absolute volume of production that underscores the importance of aquaculture, but also its further potential for delivering food protein. Aquaculture is one of the fastest-growing food sectors in the world. While annual growth rates have decreased from a peak of around 10 per cent in the 1980s and 1990s, the sector still expanded at 5.8 per cent per year from 2010 to 2016 (ibid.). This growth rate indicates the ongoing importance of aquaculture as a source of protein into the future. In 2014 aquaculture production became the world's primary source of fish food protein – with farmed fish and other aquatic organisms representing 47 per cent of the total, and 53 per cent if non-food uses (including reduction to fishmeal and fish oil) are excluded. Indeed, it appears that without the contribution of aquaculture global seafood production would not have grown. Looking to the future, total aquaculture production of food fish is expected to increase by around 30 per cent to 102 million MT by 2025 (FAO, 2016). But for the world demand for animal protein to be met in the longer term, this production will need to increase to about 140 million MT by 2050 – with the majority of production expected to be in Asia (Waite et al., 2014).

The further growth of the industry will be dependent on a range of technological transformations, including the use of feed and chemical substances and efficiencies in water use. The most significant technological gains have been made in the species most subject to international seafood trade, such as salmon, shrimp and tilapia (Kumar et al., 2018). But the future expansion of the industry will most likely be seen in a broader range of predominantly freshwater species in (but not limited to) Asia (Edwards, 2015). As outlined by Little et al. (2018), whether these species and production systems can be sustainably intensified, meeting food security needs and remaining within environmental goals, is a remaining challenge for the industry.

Trade and consumption

An enduring feature of the fisheries and aquaculture sector and the basis of the sustainable seafood movement is the global nature of seafood trade and consumption. Seafood is the most traded global food commodity in the world (Smith et al., 2010), representing about 10 per cent of total agricultural exports in terms of value. Around one-third of global seafood (worth US$143 billion) is traded internationally annually (FAO, 2018b). But as argued by Anderson et al. (2010), the seafood market is increasingly global for producers, because even if they do not export they are exposed to international competition. In fact, nearly four out of every five fish produced face international trade competition (Tveterås et al., 2012), meaning that global demand affects both the production and the price of all fish and not just the fish that are actually exported.

Trade has also increased the scope and choice of fish available to consumers. As distance has become less of a barrier to trade, wholesalers and retailers source fish from all over the world. Fish exports and imports have grown exponentially over the last 30 years, dominated by Asia and Europe. While Asia seems to have surpassed Europe as the most important seafood exporter in terms of export value (see Figures 1.2 and 1.3), Europe matches Asia in terms of import quantities and has remained the most lucrative import market in value since the early 2000s (Figures 1.4 and 1.5). In any major fish market around the world, or indeed local supermarkets, a wider selection of fish species is available than ever before. This variety is driven by the choice traders have to substitute local products with a range of international products. White fish such as cod, haddock and pollock in Europe or the US, for instance, have created demand for Nile perch from East Africa, pangasius from Vietnam, hoki from New Zealand and tilapia from Egypt (Anderson et al., 2010).

Overall, however, access to seafood in major OECD (Organisation for Economic Co-operation and Development) markets has been made possible by growing exports from the global South. The FAO notes that 'developing countries' account for 54 per cent of the total value and 59 per cent of the total exports of fish and fish products (FAO, 2018b). Demonstrating the dominant trading position of the OECD, the FAO also reports that 'developed countries' accounted for 71 per cent of global import value in both 2016 and 2017. This trade reveals the apparent dependency of these developed country markets on seafood imports, with the US and the EU importing 65 and 91 per cent of the value of domestically consumed seafood respectively, half of which comes from the global South (Pramod et al., 2014; EUMOFA, 2016).

Trade has also affected the overall availability of fish as food. Of the 171 million MT of fish produced in 2016, about 88 per cent was used for direct human consumption – a share that has increased significantly in recent decades (FAO, 2018b). Today global fish consumption is estimated at 20.5 kg

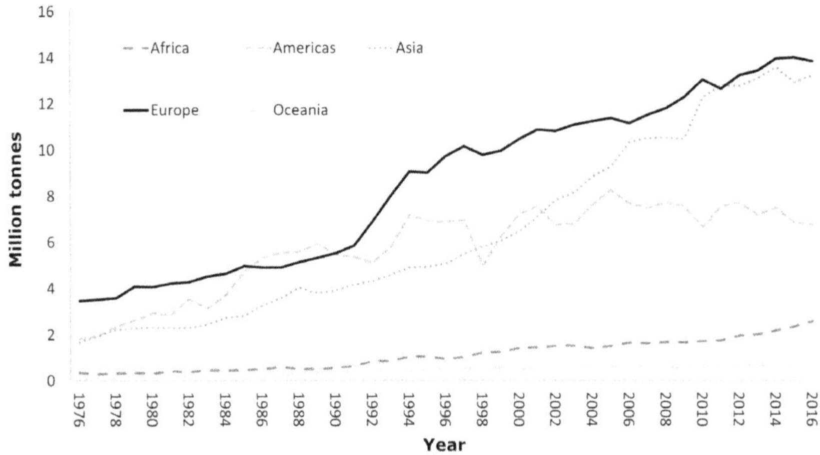

Figure 1.2 Seafood exports by continent and volume, 1976–2016.

Source: based on data from FAO (2018a).

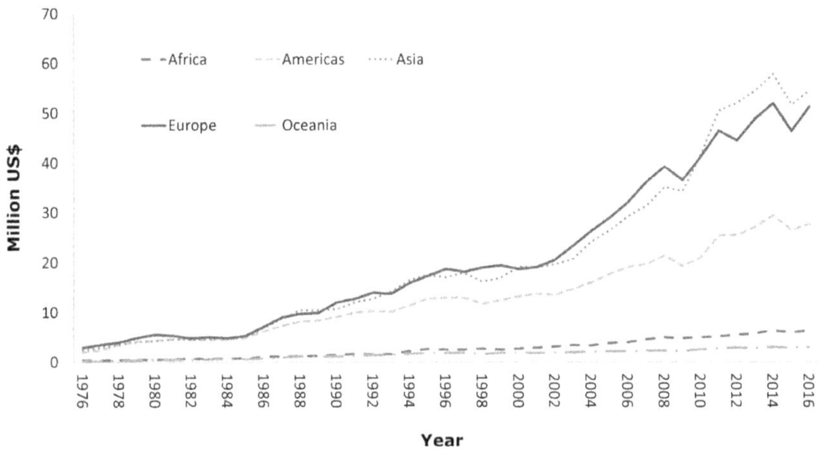

Figure 1.3 Seafood exports by continent and value, 1976–2016.

Source: based on data from FAO (2018a).

per person per year (ibid.), led in part by increased production as well as the effect of trade on production. But the relative importance of fish in human diets is also skewed. It is estimated that in 2015 seafood made up 20 per cent of the overall intake of animal protein for 3.2 billion people. But in spite of the relatively high contribution of developing countries to global trade, people in these countries have on average a higher share of fish protein in

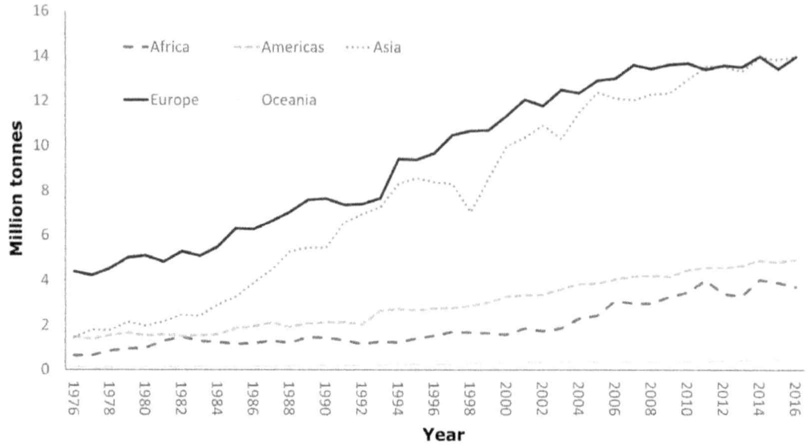

Figure 1.4 Seafood imports by continent and volume, 1976–2016 (MT).
Source: based on data from FAO (2018a).

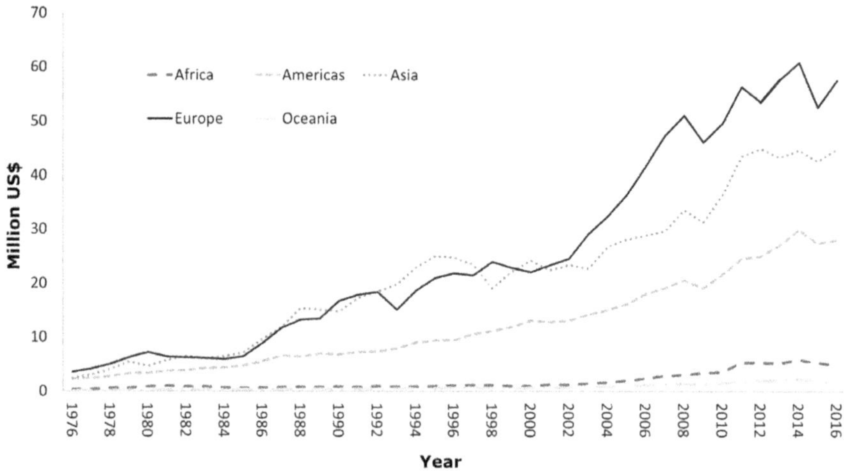

Figure 1.5 Seafood imports by continent and value, 1976–2016.
Source: based on data from FAO (2018a).

their diets than those in developed countries (ibid.). This is reflected by estimates which show that, despite their contribution to international trade, developing countries retain up to 90 per cent of the seafood they produce (Belton et al., 2017). As the middle-class urban demographic expands in these countries, it is likely that demand for more expensive fish items such as salmon and shrimp will also grow in these markets.

Over the long term trends in international seafood trade and consumption are likely to converge between developed and developing countries. Indeed, this is already happening. As reported by the FAO (2016), the 'unit value gap' between developed and developing country fish imports will continue to narrow, creating higher demand and competition for seafood globally. As outlined in this book, such trends hold consequences for the future of the sustainable seafood movement.

Sustainability problems in global seafood provision

The range of state, NGO and private sector initiatives outlined in the pages of this book reflects the broad range of sustainability issues faced in the global seafood industry. The narratives that have shaped the sustainable seafood movement have shifted over time, from capture fishers to aquaculture and from issue-specific environmental impacts to concerns over marine and coastal ecosystems, and more recently the social welfare and labour conditions of those working in the industry. Throughout this book we examine attempts to translate these issues into 'measurable' (and therefore 'verifiable') public and market claims designed to stimulate improvements in how seafood is produced, traded and consumed. The detail of these issues, measures and claims are revealed in the following chapters. Here we briefly introduce some of the key environmental, social and cross-cutting issues that underpin the sustainable seafood movement.

Fisheries' sustainability

Fisheries involve the harvest of fish and other organisms from the wild. Those engaged in fisheries are, as such, involved in a practice that is similar to any number of other extractive industries based on renewable resources. The fundamental question faced by those harvesting these organisms, and those controlling how they do so, is how much can be withdrawn before the resource exceeds its own (biological) capacity to renew itself. As outlined above, the status of fisheries is measured in exactly these terms by fishery scientists around the world. But in spite of the FAO releasing biannual status reports and national and regional management bodies constantly monitoring key input variables (such as fishing effort) and output variables (such as catches and landings), the overall sustainability of fisheries remains contested. This ongoing disagreement and uncertainty over defining and acting upon sustainability in fisheries is driven by at least three interrelated factors.

First, any assessment of stock status is complicated by the fact that we cannot directly 'see' fish in the water and instead rely on probabilistic methodologies, with levels of uncertainty dependent on the available data. The quality of available data (and the scientific capacity of stock assessors) differs considerably around the world (see for example Bartley et al., 2015; Pauly and Zeller, 2017a, 2017b; Ye et al., 2017). In some fisheries data are collected in near real time by a combination of at-sea sensing technologies tracking the location and activities of vessels and on-shore catch registration and certification. Such systems are increasingly common practice in parts of Europe, the US, some transboundary pelagic fisheries and even some small-scale coastal fisheries (e.g. Havice, 2013; Ulrich et al., 2015; Watson and Haynie, 2016; Bartholomew et al., 2018). In most countries, however, data are imprecise, inaccurate or unavailable. While small-scale coastal fisheries may have management measures in place that can compensate for weak data, larger-scale and more centralised fisheries require data as a basis for effective management.

Second, the traditional focus on the definition of biologically sustainable stocks has been expanded to include a range of ecosystem-level factors which are also subject to high levels of uncertainty. Early attention to ecosystem impacts revolved around the killing of mammals such as dolphins and turtles by using indiscriminate fishing techniques (e.g. Lewison et al., 2004). These concerns gradually evolved into a more sophisticated understanding of ecosystems and ecosystem-based fishery management, focused on the multi-species (target and non-target) nature of most fisheries and the impact of fishing techniques on benthic habitats. Today debate continues over the virtues of different modes and approaches to fishing that result from these ecosystem-level concerns. For example, is it preferable to employ balanced harvesting of fish across different trophic levels instead of targeting single species dominating single trophic levels (Garcia et al., 2012; Froese et al., 2016), or to use radical new fishing equipment such as the electrified beam trawls that minimise bycatch and impacts on benthic habitat (Haasnoot et al., 2016)?

Third, there is growing recognition that many of the input and output variables used to assess fisheries are themselves subject to a range of economic and political issues that distort any attempt to determine sustainable levels and methods of fishing. The most persistent of these issues include perverse public subsidies that support otherwise economically unfeasible fishing activity, the role of tax havens in maintaining profitability in otherwise unprofitable fisheries, exploitative access arrangements for foreign fleets in the domestic waters of (most often) developing countries and the weak enforcement of illegal fishing (see Agnew et al., 2009; Havice and Campling, 2010; Galaz et al., 2018). A basic premise surrounding these issues is that their resolution would increase the public value derived from fishery resources that could support investments in improved management. For instance, Sumaila et al. (2012) estimated that addressing key governance issues such as illegal fishing and perverse subsidies could recoup up to US$54 billion per year globally.

Aquaculture sustainability

While aquaculture has been practised in some form for centuries, if not millennia (Costa-Pierce, 1987; Nash, 2010), it is commonly seen as a new industry in many parts of the world. What is unmistakable is the huge growth in its volume, and hence the expanse of coastal areas used for raising fish and other aquatic organisms, over the last four decades. For many species this growth reached 'boom crop' proportions (Hall, 2004), with shrimp and pangasius reaching up to 35 per cent annual growth rates at their peak (FAO, 2018a). Where these kinds of growth rates have been recorded, so too have a range of sustainability concerns related to widespread conversion of land, often resulting in dramatic changes in the rural landscape and associated social transformations and environmental degradation.

Each aquaculture species faces its own set of environmental issues. For example, salmon has persistent problems of escapees (and mixing with wild salmon populations), benthic habitat fouling and lice resulting from the widespread use of sea pens (e.g. Buschmann et al., 2009), while shrimp has a range of issues related to conversion of sensitive coastal habitats, including mangrove forests, saltpans and coastal wetlands (e.g. Páez-Osuna, 2001; Cao et al., 2007). But there is also a range of issues related to good aquaculture practice that have consequences for the environment, such as the misuse of antibiotics, bycatch during the collection of wild seed and brood stock, water pollution, intensive fishing for aquaculture feeds and the introduction and transfer of non-native species (Folke and Kautsky, 1992; Pillay, 1992; Beveridge and Phillips, 1993; Boyd et al., 2005). These issues directly impact on the immediate surroundings of aquaculture systems and reduce the productivity of these systems. They also affect the social licence aquaculture holds to operate in inland and coastal areas where there are competing values and interests over the land and seascapes of which the industry is a part.

But perhaps the most persistent issue in the aquaculture industry is that of feed. The species that are mostly sold to international markets, including the US and EU, are 'non-extractive' omnivorous or carnivorous salmon, pangasius, tilapia and shrimp. The feed used to grow these species comprises fishmeal, among other ingredients, sourced from forage fish such as Peruvian anchovy. There are varying degrees of sustainability issues around the fish stocks that supply this fish – but not, it must be noted, for Peru's anchovy. In addition, the aquaculture industry has made considerable efforts to improve feed conversion ratios, so today many more salmon and shrimp can be fed with the same amount of wild fish-based feed supplemented with plant-based feed than in the past (Naylor et al., 2009). Nevertheless, 'fish as feed' remains a major issue given the broader imperatives of using fish as food for a growing global population.

Cross-cutting sustainability issues

A number of cross-cutting sustainability issues have also emerged that relate to both fisheries and aquaculture. Many of these go beyond production to incorporate different sites and practices in seafood value chains. They also go beyond environmental sustainability to include a range of social concerns as well as trade and processing practices.

Social issues have been at the forefront of seafood sustainability since being revealed by the Environmental Justice Foundation's 'Seafood not Slavefood' campaign (see EJF, 2015) and a series of high-profile exposés by investigative journalists for the *Guardian*, the *New York Times* and Associated Press (for a timeline see Bush et al., 2017). These stories highlight various aspects of unfree labour and human rights abuses from Southeast Asia to Hawaii, New Zealand, Ireland and Scotland. Labour rights abuses in the sector had been reported earlier (e.g. Derks, 2010; Resurreccion and Sajor, 2010), but these stories placed labour and the conditions under which it is employed at the centre of debates over seafood sustainability.

Trade-related issues such as fraud and food waste have also become a central focus of seafood sustainability. Several studies have revealed seafood to be one of the most commonly mislabelled international commodities, with evidence that other closely related species are regularly substituted for cod, tuna, pangasius, tilapia and shrimp (e.g. hake sold as cod) in both wholesale and retail markets (Miller and Mariani, 2010; Miller et al., 2012; Warner et al., 2013; Carvalho et al., 2017). Other studies have shown the high levels of waste in seafood value chains, with estimates of up to 37 per cent of food fish farmed or landed being wasted at distribution and retail operations and during consumption (Love et al., 2015).

As we argue later in this book, the response to these sustainability challenges has been a global effort by civil society actors and fishers alike to find alternative measures and courses of action for promoting *more* sustainable fishing under the broad umbrella of the sustainable seafood movement.

Governing seafood provision

The range of sustainability challenges facing fisheries and aquaculture emphasises the need for effective governance. This is all the more important given the highly international nature of seafood production, which requires new approaches to put pressure on regulators and producers to make the necessary changes. The result has been a gradual shift from national level to international and increasingly private forms of rule setting and enforcement.

Global fisheries' governance started with the ratification of the United Nations (UN) Convention on the Law of the Seas (UNCLOS) in 1982, which (among other fields of regulation) provided a legal framework for countries to allocate ocean territory and cooperatively manage fish stocks. But it was not until the 1992 UN Conference on Environment and Development in Rio de Janeiro that pressure for more effective fisheries' management in coastal areas led to further action. In response, the FAO developed the Code of Conduct for Responsible Fisheries (1995, containing a chapter on aquaculture) and the UN Fish Stocks Agreement (1995), both of which provided more explicit guidelines on how states could manage fish stocks. Subsequently, regional fisheries' management organisations (RFMOs) were established for the management of transboundary fish stocks. To date over 100 multilateral, regional and bilateral treaties have been agreed to supplement UNCLOS and national laws, policies and regulations. There are also 14 UN agencies and 19 international governmental organisations which have the protection of the oceans as (part of) their responsibility.

More recently (2015) the UN formulated 17 Sustainable Development Goals (SDGs) to give clear orientation to the global agenda for sustainable development. SDG 14 is devoted to marine issues: 'Conserve and sustainably use the oceans, seas and marine resources for sustainable development.'[1] Among the more specific targets, target 14.4 states that by 2020 harvesting should be effectively regulated and overfishing and illegal, unreported and unregulated (IUU) fishing should have ended. The inclusion of these goals in the global development agenda illustrates the importance of sustainability in seafood provision to the international community, and the urgency to make changes in the short term and secure its future.

But while countries sign up to international agreements, the capacity of their governments to set and implement regulation of fisheries and aquaculture has proven problematic. For instance, the weak capacity of a government to control overfishing and bycatch adequately or regulate the environmental impacts of aquaculture has been correlated with high levels of undernourishment, indicating that improved governance is essential to sustain or increase seafood's contribution to food security (Smith et al., 2010). Weak governance capacity (based on the World Bank Governance Index[2]) has also been correlated with the prevalence of illegal fishing, rather than the market price of a species or the size of a country's exclusive economic zone (EEZ) (Agnew et al., 2009). This is not to say that there are no possibilities for local-level control, often co-managed with the state. But even in the context of co-management, engagement with a competent state authority remains essential.

Global seafood governance has also developed in the field of international food safety. Established in 1963, the Codex Alimentarius (or 'Food Code') is a collection of standards, guidelines and codes of practice adopted by the Codex Alimentarius Commission under the FAO/World Health Organization Food Standards Programme.[3] The goal of the Commission and the Codex is to protect consumer health and promote fair practices in food trade. Importantly, the global architecture of the Codex provides a basis on which a range of national public and private standards and certifiers govern food safety in international trade. This architecture provides guidance for states to govern food production beyond their own borders. It also demonstrates how technical measures could be standardised by private actors in importing countries to oblige exporting countries and producers to meet internationally accepted food governance requirements.

Since the mid-1990s governmental seafood governance has been supplemented by non-governmental governance initiatives. As we explain in detail in the following chapters, these sustainability initiatives today make up the body of the sustainable seafood movement. The prime feature of these 'private' initiatives is that they respond to the perceived failures of international and national-level government initiatives to mitigate the environmental and social impacts of seafood production effectively. Central to the strategy of these initiatives, and building directly on the lessons from global food safety standards and regulation, is the market and technical logic of incentives and measurable assessments. That is, they use preferential pricing and market access to motivate improvements to production and trade practices as measured against a set of sustainability principles, indicators and criteria.

This shift from 'government to governance' to steer the seafood industry to more sustainable practice is not unique: many food sectors have experienced a similar trajectory. The key scientific and policy challenge for the sustainable seafood movement is to determine exactly how both public and private governance arrangements deliver under conditions of globalisation. We argue that certain limits have been reached in understanding and demonstrating the impact of these private governance initiatives because of their narrow focus on production, their continued technical understanding of sustainability transitions and their persistence with the logic that market demand is the major factor driving supplier improvements. We also argue that the social sciences can play an important role in overcoming these limitations and assist in developing alternative modes of seafood governance.

Seafood and social sciences

Fisheries and aquaculture have traditionally been a focus of natural science scholars, including ecologists, biologists and environmental system analysts, and resource economists. Their combined efforts have provided the basis for current technical fishery policies' focus on optimum harvest models and rights-based management (Gordon, 1954; Schaefer, 1957; Grafton et al., 2017). In aquaculture, biologists have led the development of the industry through advanced in-breeding, feed and grow-out technologies. Aquaculture economists, like their agronomist and environmental economist colleagues, have predominantly focused on production efficiency and the cost of environmental impacts (e.g. Engle, 2010; Asche and Bjorndal, 2011). The combined efforts of these scientists have been fundamental in driving technical and economics-based approaches for seafood sustainability governance.

Until the 1990s the only social scientists interested in seafood were anthropologists studying the cultural dimensions of fishing and aquaculture around the world (Symes and Hoefnagel, 2010). From a rich heritage dating back to the 1950s, anthropologists focused on the 'human dimension' of management, not only in the context of traditional fishing communities but also the consequences of attempts to implement rights-based, collective and collaborative management (McGoodwin, 1980; Acheson, 1981; Jentoft, 1989). Towards the end of the century interest from 'other' social sciences grew, including sociologists and political scientists. As Symes (2006: 114) notes, these contributions led to a 'scaling up of the focus of attention from local issues to regional and global concerns; and a move from issue-based research to the quest for a more integrated understanding of fisheries and their management' (also see Clay and McGoodwin, 1995). Social scientists started to develop theories of collaborative and cooperative forms of management between communities and the state (e.g. Jentoft et al., 1998), as well as investigating the institutional failures of resource managers to take action in the face of stock declines (Finlayson, 1994).

Reflecting the relatively young age of the aquaculture industry, 'aquacultural social science' is not as developed as fisheries' social science. Nevertheless, social scientific research has been published on the aquaculture industry since the 1980s. In the early period two distinct lines of research emerged. First, scholars were interested in the social impacts of rapidly expanding sectors, such as shrimp culture in Asia and salmon culture in Europe and North America (Bailey, 1988; Coull, 1988; Bailey et al., 1996; Barton and Staniford, 1998). Second, there was an interest in the 'history of failures in technology transfer and resistance to the regulation of natural resources' faced by the industry in Africa and Asia (Bauin, 1986; Bailey and Skladany, 1991; Harrison, 1994). In the 1990s and 2000s social scientists turned their attention to the industrial development of aquaculture. Research here focused on the emergence, organisation and social and environmental impacts of shrimp, tilapia, salmon and pangasius production (e.g. Goss et al., 2000; Phyne and Mansilla, 2003; Barton and Fløysand, 2010).

It was not until the turn of the millennium that social scientists started to set a new agenda around fisheries' and aquaculture 'governance' (McCay, 2000; Symes and Hoefnagel, 2010). Small-scale fisheries and fishery co-management featured heavily in this work (Jentoft et al., 1998). However, the focus gradually expanded as the intensification of global connections and interactions in seafood provision expanded the number of stakeholders that need to engage in the control over these resources to include NGOs and the private sector. Various theoretical approaches to fisheries' governance have been developed in recognition of this wider group of actors and issues, including adaptive governance (Folke et al., 2005) and interactive fisheries' governance (Kooiman et al., 2005). As we outline in more detail in Chapter 2, it is only recently that the social sciences have expanded to questions of more networked global governance arrangements, such as the legitimacy of labelling and certification, the interaction between public and private rule making and the role of consumers and retailers.

This book contributes to the sociology of globalisation by focusing on the social relations and networks that condition the exploitation and sustainability of fisheries and aquaculture production. We set a new agenda for a seafood social science that explores the complex ways in which different societal actors engage across global space to solve complex environmental and social governance challenges. It also allows social scientists to contribute a social understanding to supplement the predominantly technical and economic perspectives dominating seafood science and policy. To achieve this we advocate for the greater engagement of social sciences of all breeds to engage with and research the practices of producers, policy practitioners and private sector actors who are central to the sustainability outcomes of the seafood industry (and indeed any global industry). Only then, we argue, will social scientists have a permanent seat at the table, and in doing so move beyond what Symes and Hoefnagel (2010) argue has been a largely incidental and isolated position in fishery (and aquaculture) policy to date.

Outline of the book

The rest of this book is divided into seven chapters grouped into three parts. Part I (Chapters 2 and 3) provides an overview of the empirical and conceptual issues relevant for understanding the governance of sustainable seafood. Part II (Chapters 4–6) presents different ways of regulating seafood around the world. Part III (Chapters 7 and 8) looks ahead to the future of sustainable seafood.

Chapter 2 expands on a social scientific understanding of the sustainable seafood movement. We first review debates on governance, including the role of public and private legitimacy and authority before turning to global value chain (GVC) analysis as a common framework for understanding the organisation and interaction of key public and private actors. In light of the limitations of the GVC framework we elaborate on the potential for an alternative framework based on theories of flows and networks and social movements. This framework provides the basis for identifying the challenges facing global seafood provision and the governance responses from public and private actors in the following chapters. It also establishes a basis for further reflection on an alternative conceptualisation of the sustainable seafood movement introduced in Chapter 7.

Chapter 3 provides an overview of the global scale of seafood production, trade and consumption, and dives into geographical differences across the world. Starting with a general description of major statistical trends over the last quarter of a century, the chapter outlines the current status of global seafood production and consumption. The chapter then turns to a general description of the rise of the 'global' sustainable seafood movement, and explores how it has influenced (the governance of) seafood production and consumption practices in different parts of the world. For example, what impact does an ostensibly Anglo-American movement have on Asia, the largest seafood-producing region in the world? What differences exist even within and between North America and Europe? What is the relevance of different scales in which fisheries are organised in different parts of the world? The final section turns to the implications these geographies might have on state and market governance arrangements.

Chapter 4 outlines the role of public regulation of seafood production and consumption. A primary assumption of the sustainable seafood movement is the perceived failure of the state to regulate fisheries and aquaculture production effectively. The chapter investigates whether this is a fair assessment of the role of the state in seafood governance by outlining the experience of governments acting either within their own territories or collectively in multilateral arrangements in regional seas and high-seas areas. Are these perceptions of failure justified? What differences exist between fisheries and aquaculture? And what alternatives are there? The chapter goes on to explore the interaction between the normative goals of sustainability and trade dominated by Northern over Southern states via the World Trade Organization (WTO) as well as bilaterally.

Chapter 5 reviews the wide range of market-based approaches to sustainability governance that have gained considerable traction in the seafood movement. The chapter provides an overview of these approaches in fisheries and aquaculture, including but not limited to NGO-led initiatives such as the MSC and the Aquaculture Stewardship Council (ASC), and industry-led initiatives such as GlobalGAP, the Global Aquaculture Alliance (GAA) and the Responsible Fishing Scheme. We discuss how these arrangements have led to the inclusion of new categories of private actors in the regulation of sustainable seafood (e.g. auditors), and a new round of contestation between NGOs and industry actors such as retailers, which are seeking to reduce their reputational risk with a series of benchmarking exercises and internal certification schemes. The chapter concludes with a discussion of whether and how the investment made in these market approaches has supplemented (or even replaced) state governance arrangements in promoting sustainable seafood production and consumption.

Chapter 6, the last of the empirical chapters, goes beyond the discussion of state and/or market governance arrangements to identify a range of alternative, often niche, approaches to promoting seafood sustainability. Here we focus on how these alternative approaches constitute a range of innovative governance arrangements that draw out the importance of understanding the role of social dynamics in determining the sustainability of seafood production and trade. A number of examples are presented, including locally supported fisheries, pledges for sustainable seafood, the Slow Fish movement, the role of chefs in promoting sustainable seafood and organic production. In short, we consider how various market-based initiatives address some of the challenges faced by the mainstream approaches outlined in the previous chapters. What is the potential impact of such initiatives on promoting sustainable seafood? And, responding to Chapter 3, do they create new geographies of sustainable production and consumption that include Asia and the global South?

In Chapter 7 we argue for a fundamentally social approach to understanding the global social relations that condition the success of the sustainable seafood movement. We first reflect on the empirical chapters and the relationship between the sustainability initiatives under the sustainable seafood movement; these are linked to themes of global/local, inclusion/exclusion, public/private and convergence/divergence. We then return to the theory proposed in Chapter 2 to develop a new flows and practices perspective for understanding global food movements.

In the final chapter we present the case for an integrated approach to seafood governance that draws on state and market governance arrangements to evaluate and act upon the international social relations of production and consumption that ultimately determine sustainable seafood practices. Emphasising these social relations, we argue, can move beyond the narrow focus of technocratic evaluations and economic incentives for promoting seafood to enable new pathways to sustainable production, trade and consumption. To this end, we close the book with seven principles for implementing an integrated governance approach to seafood and a starting point for reassessing the role and impact of social, state and market governance initiatives.

Notes

1 See www.un.org/sustainabledevelopment/sustainable-development-goals/.
2 The World Bank Governance Index is based on six indicators: voice and accountability, political stability and absence of violence, government effectiveness, regulatory quality, rule of law and control of corruption
3 See the website of the Codex Alimentarius for details on the history and range of standards under its control: www.fao.org/fao-who-codexalimentarius/about-codex/history/en/.

References

Acheson, J. M. (1981). 'Anthropology of fishing', *Annual Review of Anthropology* 10(1), 275–316.

Agnew, D. J., J. Pearce, G. Pramod, T. Peatman, R. Watson, J. R. Beddington and T. J. Pitcher (2009). 'Estimating the worldwide extent of illegal fishing', *PLoS One* 4(2), e4570.

Anderson, J. L., F. Asche and S. Tveterås (2010). 'World fish markets', in R. Q. Grafton, H. Hilborn, D. Squires, M. Tait and M. J. Williams (eds) *Handbook of Marine Fisheries Conservation and Management.* Oxford: Oxford University Press, 113–123.

Asche, F. and T. Bjorndal (2011). *The Economics of Salmon Aquaculture.* Chichester: John Wiley & Sons.

Bailey, C. (1988). 'The social consequences of tropical shrimp mariculture development', *Ocean and Coastal Management* 11: 31–44.

Bailey, C., S. Jentoft and P. Sinclair (1996). 'Social sciences contributions to aquaculture development', in C. Bailey, S. Jentoft and P. Sinclair (eds) *Aquaculture Development: Social Dimensions of an Emerging Industry.* Boulder, CO: Westview Press, 3–22.

Bailey, C. and M. Skladany (1991). 'Aquaculture development in tropical Asia', *Natural Resources Forum* 15(66–73).

Bartholomew, D. C., J. C. Mangel, J. Alfaro-Shigueto, S. Pingo, A. Jimenez and B. J. Godley (2018). 'Remote electronic monitoring as a potential alternative to on-board observers in small-scale fisheries', *Biological Conservation* 219, 35–45.

Bartley, D. M., G. J. De Graaf, J. Valbo-Jørgensen and G. Marmulla (2015). 'Inland capture fisheries: Status and data issues', *Fisheries Management and Ecology* 22(1), 71–77.

Barton, J. R. and A. Fløysand (2010). 'The political ecology of Chilean salmon aquaculture, 1982–2010: A trajectory from economic development to global sustainability', *Global Environmental Change* 20(4), 739–752.

Barton, J. R. and D. Staniford (1998). 'Net deficits and the case for aquacultural geography', *Area* 30(2), 145–155.

Bauin, S. (1986). 'Aquaculture: A field by bureaucratic fiat', in M. Callon, J. Law and A. Rip (eds) *Mapping the Dynamics of Science and Technology*. London: Macmillan, 124–141.

Belton, B., S. R. Bush and D. Little (2017). 'Not just for the wealthy: Rethinking farmed fish consumption in the global South', *Global Food Security* 16, 85–92.

Beveridge, M. and M. Phillips (1993). 'Environmental impact of tropical inland aquaculture', *Environment and Aquaculture in Developing Countries* 31, 213–236.

Boyd, C. E., A. A. McNevin, J. Clay and H. M. Johnson (2005). 'Certification issues for some common aquaculture species', *Reviews in Fisheries Science* 13(4), 231–279.

Buschmann, A. H., F. Cabello, K. Young, J. Carvajal, D. A. Varela and L. Henríquez (2009). 'Salmon aquaculture and coastal ecosystem health in Chile: Analysis of regulations, environmental impacts and bioremediation systems', *Ocean & Coastal Management* 52(5), 243–249.

Bush, S. R., M. J. Marschke and B. Belton (2017). 'Labor, social sustainability and the underlying vulnerabilities of work in Southeast Asia's seafood value chains', in A. McGregor, L. Law and F. Miller (eds) *Routledge Handbook of Southeast Asian Development*. London: Routledge, 316–329.

Bush, S., H. Toonen, P. Oosterveer and A. Mol (2013). 'The "devil's triangle" of MSC certification: Balancing credibility, accessibility and continuos improvement', *Marine Policy* 37, 288–293.

Cao, L., W. Wang, Y. Yang, C. Yang, Z. Yuan, S. Xiong and J. Diana (2007). 'Environmental impact of aquaculture and countermeasures to aquaculture pollution in China', *Environmental Science and Pollution Research* 14(7), 452–462.

Carvalho, D. C., D. Guedes, M. da Gloria Trindade, R. M. S. Coelho and P. H. de Lima Araujo (2017). 'Nationwide Brazilian governmental forensic programme reveals seafood mislabelling trends and rates using DNA barcoding', *Fisheries Research* 191, 30–35.

Clay, P. M. and J. R. McGoodwin (1995). 'Utilizing social sciences in fisheries management', *Aquatic Living Resources* 8, 203–207.

Costa-Pierce, B. A. (1987). Aquaculture in ancient Hawaii. *BioScience* 37(5): 320–331.

Costello, C., D. Ovando, R. Hilborn, S. D. Gaines, O. Deschenes and S. E. Lester (2012). 'Status and solutions for the world's unassessed fisheries', *Science* 1224768.

Coull, J. R. (1988). 'Fish farming in the Highlands and Islands: Boom industry of the 1980s', *Scottish Geographical Magazine* 104(1), 4–13.

Derks, A. (2010). 'Migrant labour and the politics of immobilisation: Cambodian fishermen in Thailand', *Asian Journal of Social Science* 38(6), 915–932.

Edwards, P. (2015). 'Aquaculture environment interactions: Past, present and likely future trends', *Aquaculture* 447, 2–14.

Edwards, P. and H. Demaine (1998). *Rural Aquaculture: Overview and Framework for Country Reviews*. Bangkok: FAO.

EJF (2015). *Thailand's Seafood Slaves. Human Trafficking, Slavery and Murder in Kantang's Fishing Industry*. London: EJF.

Engle, C. R. (2010). *Aquaculture Economics and Financing: Management and Analysis.* Ames, IA: John Wiley & Sons.

EUMOFA (2016). *The EU Fish Market.* Brussels: European Market Observatory for Fisheries and Aquaculture Products.

FAO (2016). *The State of World Fisheries and Aquaculture.* Rome: FAO Fisheries and Aquaculture Department.

FAO (2018a). 'FAOSTAT', FAO, available at http://faostat.fao.org/faostat/ (accessed 23 August 2018).

FAO (2018b). *The State of World Fisheries and Aquaculture.* Rome: FAO Fisheries and Aquaculture Department.

Finlayson, A. C. (1994). *Fishing for Truth: A Sociological Analysis of Northern Cod Stock Assessments from 1977 to 1990.* St. John's, NL: Institute of Social and Economic Research, Memorial University of Newfoundland St. John's.

Fletcher, W. J., J. Chesson, M. Fisher, K. J. Sainsbury, T. Hundloe, A. D. M. Smith and B. Whitworth (2002). *National ESD Reporting Framework for Australian Fisheries: The 'How To' Guide for Wild Capture Fisheries.* Canberra, CT: FRDC Project 2000/145.

Folke, C. and N. Kautsky (1992). 'Aquaculture with its environment: Prospects for sustainability', *Ocean & Coastal Management* 17(1), 5–24.

Folke, C., T. Hahn, P. Olsson and J. Norberg (2005). 'Adaptive governance of social-ecological systems', *Annual Review of Environmental Resources* 30, 441–473.

Froese, R., C. Walters, D. Pauly, H. Winker, O. L. F. Weyl, N. Demirel, A. C. Tsikliras and S. J. Holt (2016). 'A critique of the balanced harvesting approach to fishing', *ICES Journal of Marine Science* 73(6), 1640–1650.

Galaz, V., B. Crona, A. Dauriach, J.-B. Jouffray, H. Österblom and J. Fichtner (2018). 'Tax havens and global environmental degradation', *Nature Ecology & Evolution* 2, 1352–1357.

Garcia, S., J. Kolding, J. Rice, M.-J. Rochet, S. Zhou, T. Arimoto, J. Beyer, L. Borges, A. Bundy and D. Dunn (2012). 'Reconsidering the consequences of selective fisheries', *Science* 335(6072), 1045–1047.

Gordon, H. S. (1954). 'The economic theory of a common-property resource: The fishery', in *Classic Papers in Natural Resource Economics.* Springer, 178–203.

Goss, J., D. Burch and R. Rickson (2000). 'Agri-food restructuring and third world transnationals: Thailand, the CP Group and the global shrimp industry', *World Development* 28(3), 513–530.

Grafton, R. Q., J. Kirkley and D. Squires (2017). *Economics for Fisheries Management.* London, Routledge.

Haasnoot, T., M. Kraan and S. R. Bush (2016). 'Fishing gear transitions: Lessons from the Dutch flatfish pulse trawl', *ICES Journal of Marine Science* 73(4), 1235–1243.

Hall, D. (2004). 'Explaining the diversity of Southeast Asian shrimp aquaculture', *Journal of Agrarian Change* 4(33), 315–335.

Harrison, E. (1994). 'Aquaculture in Africa: Socio-economic dimensions', in J. F. Muir and R. J. Roberts (eds) *Recent Advances in Aquaculture V.* London: Croom Helm, 240–299.

Hatanaka, M., C. Bain and L. Busch (2005). 'Third-party certification in the global agrifood system', *Food Policy* 30, 354–369.

Havice, E. (2013). 'Rights-based management in the Western and Central Pacific Ocean tuna fishery: Economic and environmental change under the Vessel Day scheme', *Marine Policy* 42, 259–267.

Havice, E. and L. Campling (2010). 'Shifting tides in the Western and Central Pacific Ocean tuna fishery: The political economy of regulation and industry responses', *Global Environmental Politics* 10(1), 89–114.

Jacquet, J. L. and D. Pauly (2007). 'The rise of seafood awareness campaigns in an era of collapsing fisheries', *Marine Policy* 31(3), 308–313.

Jentoft, S. (1989). 'Fisheries co-management: Delegating government responsibility to fishermen's organizations', *Marine Policy* 13(2), 137–154.

Jentoft, S., B. J. McCay and D. C. Wilson (1998). 'Social theory and fisheries co-management', *Marine Policy* 22(4–5), 423–436.

Kooiman, J., S. Jentoft, M. Bavinck and R. Pullin (eds) (2005). *Fish for Life: Interactive Governance for Fisheries*. Amsterdam: Amsterdam University Press.

Kumar, G., C. Engle and C. Tucker (2018). 'Factors driving aquaculture technology adoption', *Journal of the World Aquaculture Society* 49(3), 447–476.

Lewison, R. L., L. B. Crowder, A. J. Read and S. A. Freeman (2004). 'Understanding impacts of fisheries bycatch on marine megafauna', *Trends in Ecology & Evolution* 19(11), 598–604.

Little, D. C., J. A. Young, W. Zhang, R. W. Newton, A. Al Mamun and F. J. Murray (2018). 'Sustainable intensification of aquaculture value chains between Asia and Europe: A framework for understanding impacts and challenges', *Aquaculture* 493, 338–354.

Love, D. C., J. P. Fry, M. C. Milli and R. A. Neff (2015). 'Wasted seafood in the United States: Quantifying loss from production to consumption and moving toward solutions', *Global Environmental Change* 35, 116–124.

McCay, B. J. (2000). 'Sea changes in fisheries policy: Contributions from anthropology', in Paul E. Durrenberger, and Thomas D. King (eds) *State and Community in Fisheries Management: Power, Policy and Practice*. Westport, CT: Bergin & Garvey, 201–217.

McGoodwin, J. R. (1980). 'Mexico's marginal inshore Pacific fishing cooperatives', *Anthropological Quarterly* 53(1), 39–47.

Miller, D., A. Jessel and S. Mariani (2012). 'Seafood mislabelling: Comparisons of two Western European case studies assist in defining influencing factors, mechanisms and motives', *Fish and Fisheries* 13(3), 345–358.

Miller, D. D. and S. Mariani (2010). 'Smoke, mirrors, and mislabeled cod: Poor transparency in the European seafood industry', *Frontiers in Ecology and the Environment* 8(10), 517–521.

Nash, C. (2010). *The History of Aquaculture*. Ames, IA: John Wiley & Sons.

Naylor, R. L., R. W. Hardy, D. P. Bureau, A. Chiu, M. Elliott, A. P. Farrell, I. Forster, D. M. Gatlin, R. J. Goldburg, K. Hua and P. D. Nichols (2009). 'Feeding aquaculture in an era of finite resources', *Proceedings of the National Academy of Sciences* 106(36), 15103–15110.

Páez-Osuna, F. (2001). 'The environmental impact of shrimp aquaculture: Causes, effects, and mitigating alternatives', *Environmental Management* 28(1), 131–140.

Parkes, G., S. Walmsley, T. Cambridge, R. Trumble, S. Clarke, D. Lamberts, D. Souter and C. White (2009). 'Review of Fish Sustainability Information Schemes; Final Report.' Edinburgh: MRAG and Stirling University.

Pauly, D. and D. Zeller (2017a). 'The best catch data that can possibly be? Rejoinder to Ye et al. "FAO's statistic data and sustainability of fisheries and aquaculture"', *Marine Policy* 81, 406–410.

Pauly, D. and D. Zeller (2017b). 'Comments on FAO's State of World Fisheries and Aquaculture (SOFIA 2016)', *Marine Policy* 77, 176–181.

Phyne, J. and J. Mansilla (2003). 'Forging linkages in the commodity chain: The case of the Chilean salmon farming industry, 1987–2001', *Sociologia Ruralis* 43(2), 108–127.

Pillay, T. V. R. (1992). *Aquaculture and the Environment*. Oxford: Fishing News Books.

Pramod, G., K. Nakamura, T. J. Pitcher and L. Delagran (2014). 'Estimates of illegal and unreported fish in seafood imports to the USA', *Marine Policy* 48, 102–113.

Resurreccion, B. P. and E. E. Sajor (2010). '"Not a real worker": Gendering migrants in Thailand's shrimp farms', *International Migration* 48(6), 102–131.

Roheim, C. A. (2009). 'An evaluation of sustainable seafood guides: Implications for environmental groups and the seafood industry', *Marine Resource Economics* 24, 301–310.

Roheim, C. A., S. R. Bush, F. Asche, J. N. Sanchirico and H. Uchida (2018). 'Evolution and future of the sustainable seafood market', *Nature Sustainability* 1(8), 392–398.

Safina, C. (1995). 'The world's imperiled fish', *Scientific American* 273, 46–53.

Schaefer, M. B. (1957). 'Some considerations of population dynamics and economics in relation to the management of the commercial marine fisheries', *Journal of the Fisheries Board of Canada* 14(5), 669–681.

Smith, M. D., C. A. Roheim, L. B. Crowder, B. S. Halpern, M. Turnipseed, J. L. Anderson, F. Asche, L. Bourillón, A. G. Guttormsen and A. Khan (2010). 'Sustainability and global seafood', *Science* 327(5967), 784–786.

Sumaila, U. R., W. Cheung, A. Dyck, K. Gueye, L. Huang, V. Lam, D. Pauly, T. Srinivasan, W. Swartz, R. Watson and D. Zeller (2012). 'Benefits of rebuilding global marine fisheries outweigh costs', *PLoS One* 7(7), e40542.

Symes, D. (2006). 'Fisheries governance: A coming of age for fisheries social science?', *Fisheries Research* 81(2), 113–117.

Symes, D. and E. Hoefnagel (2010). 'Fisheries policy, research and the social sciences in Europe: Challenges for the 21st century', *Marine Policy* 34(2), 268–275.

Tveterås, S., F. Asche, M. F. Bellemare, M. D. Smith, A. G. Guttormsen, A. Lem, K. Lien and S. Vannuccini (2012). 'Fish is food – The FAO's fish price index', *PLoS One* 7(5), e36731.

Ulrich, C., H. J. Olesen, H. Bergsson, J. Egekvist, K. B. Håkansson, J. Dalskov, L. Kindt-Larsen and M. Storr-Paulsen (2015). 'Discarding of cod in the Danish Fully Documented Fisheries trials', *ICES Journal of Marine Science* 72(6), 1848–1860.

Waite, R., M. Beveridge, R. Brummett, S. Castine, N. Chaiyawannakarn, S. Kaushik, R. Mungkung, S. Nawapakpilai and M. Phillips (2014). *Improving Productivity and Environmental Performance of Aquaculture*. Penang: WorldFish.

Warner, K., W. Timme, B. Lowell and M. Hirschfield (2013). 'Oceana study reveals seafood fraud nationwide.' Washington, DC: Oceana.

Watson, J. T. and A. C. Haynie (2016). 'Using vessel monitoring system data to identify and characterize trips made by fishing vessels in the United States North Pacific', *PLoS One* 11(10), e0165173.

Worm, B., R. Hilborn, J. K. Baum, T. A. Branch, J. S. Collie, C. Costello, M. J. Fogarty, E. A. Fulton, J. A. Hutchings, S. Jennings, O. P. Jensen, H. K. Lotze, P. M. Mace, T. R. McClanahan, C. Minto, S. R. Palumbi, A. M. Parma, D. Ricard, A. A. Rosenberg, R. Watson and D. Zeller (2009). 'Rebuilding global fisheries', *Science* 325(5940), 578–585.

Ye, Y., M. Barange, M. Beveridge, L. Garibaldi, N. Gutierrez, A. Anganuzzi and M. Taconet (2017). 'FAO's statistic data and sustainability of fisheries and aquaculture: Comments on Pauly and Zeller (2017)', *Marine Policy* 81, 401–405.

Part I

Understanding sustainable seafood

2 Understanding sustainable global seafood governance

Introduction

A rich and diverse set of theoretical approaches is used to analyse the governance of global food provisioning. In this chapter we review these approaches and assess their relevance for understanding the challenges, limitations and innovation observed in the range of governance initiatives that fall under the umbrella of the sustainable seafood movement.

The chapter introduces the concept of governance and how it helps to identify the emergence, legitimacy and authority of new hybrid state–market–civil society arrangements. We present main concepts around the GVC framework, reviewing both its strengths and its weaknesses as a framework for understanding the organisation and influence of public and private actors in the global seafood industry. Recognising the limitations of the GVC framework, we then elaborate an alternative framework based on theories of flows and networks and social movements. We argue that this combination of theories provides a stronger basis for an alternative conceptualisation of the sustainable seafood movement.

From government to governance in managing seafood provision

What is this thing called governance?

The seafood sector presents an illustrative example of the shift from 'government to governance' in response to transformation of supply chains across global space. Seafood is produced using many different practices under very different politico-administrative conditions, all of which are connected through global supply chains. State regulatory systems have therefore been replaced by more diverse regulatory arrangements involving different private and non-governmental actors, which act in varying degrees of partnership with conventional governments (Pattberg, 2007; Gulbrandsen, 2012). As a result, contemporary sustainable seafood governance has become a complex mosaic of public and private governance instruments and arrangements which connect different locations, scales, actors and sustainability concerns. A short review of the fisheries and aquaculture sectors highlights this shift to governance.

Until the establishment of the 200-mile EEZs in 1977, the oceans constituted a common resource to which all nations had equal fishing rights. Still today, the high seas beyond the EEZs are not under the formal control of governmental authorities, and this means that around 64 per cent of the oceans lie beyond the jurisdiction of any one country. Increasing pressures on marine fish stocks and diminishing resources have shown that this conventional approach to fisheries' management risks is becoming ineffective. As early as the 1992 UN Conference on Environment and Development, participants concluded that 'the traditional regime of freedom of the seas' should be transformed into a regime that is characterised 'by more purposeful and detailed management' at a transnational scale (Juda, 2002: 128). As outlined by Hoel et al. (2005), the evident impacts of overfishing and environmental degradation have forced a shift in fisheries' policies away from a focus on economic optimisation to one on the promotion of sustainable management of the remaining marine resources.

Although the need for sustainable seafood management is widely accepted, developing and implementing an effective system under the conditions of globalisation have proven complicated. For fisheries, setting sustainable levels of fishing pressure for the anchovy fisheries off the coast of Peru is influenced in large part by shifting demand for fishmeal and fish oil in the global animal feed sector. Establishing local rights for fishers off the coast of West Africa is determined in part by international trade agreements with the EU. Sustainable shrimp aquaculture is linked to local land and water management, as are the management of feed inputs ranging from anchovy fisheries in Peru and soybean agriculture in Brazil. The ability of producers to respond to sustainability issues is also determined by changing consumer demand and import tariffs. This picture of complexity highlights the challenge for any single state to manage the seafood within its own jurisdiction.

These increasing interdependencies and complexities in global seafood supply chains, coupled with the rise of complex sustainability challenges, have driven the search for governance arrangements that can transcend the nation-state.

Governance coordinates and organises societal responses to (global) problems by fostering collective determination of the fundamental values and principles for management arrangements (Chuenpagdee and Jentoft, 2009). Those involved in governance initiatives question the norms and values that underlie societal agreement over what constitutes 'effective' control and even 'acceptable' outcomes for complex issues, and commonly seek new modes of civic action and coordination. For states this means sharing the responsibility of defining these norms and values with both supply-chain actors (Marsden et al., 2010) and civil society organisations (Hatanaka et al., 2005; Boström and Klintman, 2008). The result is that hierarchical, state-led and coordinated forms of environmental regulation and control are giving way to more flexible and network-based modes of environmental governance that are better able to deal with the many interdependencies that exist within and between societal groups (Mohammed, 2014).

Central to the these new societal interdependences are networks, which, as outlined by Fulponi (2006), are observed in three ways. First, networked interactions have led to the emergence of private voluntary systems of governance which address process as well as product characteristics. Second, networds foster the continual reproduction of global coalitions for setting principles, standards and assessment methodologies. And third, they set new conditions for interaction between businesses engaged in the production, trade and retail of food products. As we now discuss, these trends are also visible in the global seafood sector.

Transitions in global seafood governance

Contemporary governance of seafood sustainability is in many ways a complex microcosm of innovation when it comes to understanding the changing roles and interactions between public and private actors. The shifting role of the state in the fisheries' sector offers a particular set of insights into the changing role of nation-states, for the very reason that fish stocks are public resources. Aquaculture also offers new insights into the changing role of governments because of the often ambiguous tenure over coastal and riparian lands used for fish ponds and cages, and the fundamental reliance on (what are in the majority of countries) public goods like water and fishmeal for production. As variously noted (Held et al., 1999; Konefal et al., 2005), the governance of seafood is therefore fundamentally about power sharing and negotiated decision-making over the use of these public resources, and the inclusion of private non-state interests such as retailers, traders, farmers, fishers, consumers and NGOs.

But the involvement of these actors does not mean there has been a wholesale shift from state to non-state or public to private governance arrangements. Instead, based on Imperial and Yandle (2005) and Oosterveer and Sonnenfeld (2012), we identify four broad categories of seafood governance approaches that co-exist and interact.

First, we see *centralised bureaucratic arrangements* involving government regulation based on sovereign control over territorial waters and regulatory control over the 200 nautical mile EEZs and the formation of RFMOs under the 1995 UN Transboundary Fish Stocks Agreement. The basis of centralised systems of state control is the spatial enclosure of public property. The premise of establishing clear property rights over fish stocks in the EEZs was to extend regulatory power over these resources (DeSombre and Barkin, 2011; Campling and Havice, 2014). However, the role of states is more complicated – and often contradictory. They often subsidise the development of fishing fleets, leading to overexploitation of resources (e.g. Kelleher et al.,

2009; Sala et al., 2018), while at the same time trying to maintain a degree of monitoring and assessment of the status of their fish stocks – most notably through the development of key stock indicators such as maximum sustainable yield (Finley, 2011). At the RFMO level a similar pattern is observable, with states seeking to negotiate multilateral access and harvest control rules over fish stocks in the high seas.

Second, states have adopted a range of *market-based arrangements* to allocate the available resources among private citizens who are allowed to trade their quota. Faced with the expansion of effort in the EEZs, and later in the high seas, states have sought mechanisms to allocate and control fishing effort. For instance, setting and trading individual transferable quotas (ITQs) or licences assign shares in a fishery resource within a total amount of allowable catch or units of effort, or designate exclusive geographical areas and time periods for fishing to individual fishers. By making these 'rights' tradable, economic pressure is placed on individuals to increase their efficiency by innovating fishing practices (Constanza et al., 1999; Barkin and DeSombre, 2013). A similar but far less well-developed system is the use of environmental credit systems, aimed at incentivising a greater share of the catch in return for investments in changed fishing gear or practices (Van Riel et al., 2015). While controversial, these approaches remain relatively mainstream – based on the common understanding that economic incentives can drive fishers to achieve sustainability goals.

Third, *civil-society-based arrangements* use a wide range of standards, codes, labelling and certification schemes to set normative rules for regulating production and trade. The range of these initiatives is impressive, varying from ecolabels such as the MSC to consumer boycotts and remote monitoring systems on illegal fishing (Iles, 2007; Jacquet and Pauly, 2007). These strategies build on moral notions of 'doing the right thing' and cognitive notions of not doing things that are contradictory to prevailing norms or knowledge (Cashore, 2002). By using these tools, civil society actors have been able to create institutional bridges through monitoring, advocacy and lobbying. In doing so these actors engage with industry actors to shift or internalise an environmental rationality for improved practices with the prevailing market logic of economic gain (Oosterveer and Spaargaren, 2011). At the same time, these civil society actors seek to maintain their advocacy function by continuing to engage with governments in maintaining or improving rules for production.

Fourth, there has been a significant rise in *self-regulation arrangements* by communities and resource users. Falling under the broader category of community-based resource management, these arrangements seek to counter centralised forms of regulation. Resource users under these arrangements organise resource use according to customary norms and rules, or under norms and rules set out by new organisations such as producer associations (Jentoft, 2000; Kooiman and Bavinck, 2005). These norms and rules can regulate both access to the resource and fishing practices, based on a wide array of justifications related to social cohesion, religion or management guidelines based on local ecological knowledge. In many if not all cases there is a degree of state involvement in various modes of co-management arrangements, ranging from a minimal advisory role to being directly involved in the formulation and implementation of management rules and arrangements (Sen and Raakjaer Nielsen, 1996).

While the form and function of these different arrangements have been examined in specific cases, countries and regions, less attention has been given how such arrangements interact under conditions of globalisation. In particular, how does the global nature of the sustainable seafood movement influence the emergence, design and interaction of these public and private forms of governance? Before outlining a framework for addressing this question, we first explore the tensions and challenges confronting these networked forms of governance.

Selected themes in governance research

Legitimacy in global sustainability governance

To be effective, non-state or private governance initiatives must be legitimate in the eyes of those they seek to regulate. But with the formal procedures of government, private arrangements have to generate their legitimacy by constantly (re)producing the trust of other societal actors. The legitimacy of private initiatives is as such relational and dynamic, subject to 'the perception and assumption that (their) operations are trustworthy, responsible, desirable and appropriate' (Boström, 2006: 351). If a private set of rules, and the actors that develop them, are deemed legitimate by those that are to be governed, these rules are afforded the authority to implement conservation activities (Cashore et al., 2004).

The legitimacy of private standards is generated through three different but related processes (Fuchs et al., 2011). *Input* legitimacy refers to the extent to which stakeholders are included in the decision-making process: setting the agenda, selecting the issues involved and therefore influencing the outcomes (Rantala, 2012; Eshuis and Edwards, 2013). *Throughput* legitimacy relates to the quality of the decision-making process throughout the lifetime of a conservation activity, including the transparency, responsiveness and fairness of related rules and norms (Fuchs et al., 2011). Finally, *output* legitimacy relates to the extent to which sustainability outcomes from private-led conservation initiatives are observable in terms of the degree to which their goals are reached. In other words, does the private initiative make a difference? A critical dimension of the legitimacy of private forms of governance is their degree of transparency. Given the relational nature of legitimacy, these governance arrangements have to secure openness to create trust between the various actors involved in highly (spatially) dispersed networks. Their ability to do this is fundamental to their overall effectiveness.

There are also several critical issues challenging the legitimacy of private governance initiatives. First, private governance arrangements may be biased by involving selected categories of stakeholders and addressing only particular sustainability concerns while ignoring broader injustices. Others observe a process of 'crowding out' (Vandergeest, 2007), where selective and biased private governance arrangements crowd out more stringent and more encompassing government interventions. Second, stakeholders in developing countries may resist private certification as a (green) imperialist form of extraterritoriality (Vandergeest and Unno, 2012) wherein Western consumers, governments and large private companies impose their concerns upon smallholder producers (Constance and Bonanno, 1999, 2000; Campbell, 2005; Casey, 2008; Cheyns, 2011). Third, authority may be granted to private initiatives although they lack apparent legitimacy because the standard has acquired dominance in the market (Miller and Bush, 2015). Together, these critical issues point to the precarious nature of legitimacy in private global seafood governance and the need to consider it when assessing the overall effectiveness of private initiatives.

Public–private interaction

The interaction between public and private regulation is a key determinant of governance success. Such interaction was originally thought to work in one of two directions. First, by setting 'higher' standards than state legislation, private governance was believed to incentivise governments to 'ratchet up' their regulatory performance (Cashore et al., 2007). Second, where private governance was thought to have emerged in response to overregulation by states, setting minimum industry requirements to avoid regulatory inflation, the outcome was widely considered to be a race to the bottom as firms bargain with what constitutes minimum performance (Nadvi and Wältring, 2004). But today the interaction between public and private governance is understood to be far more complex, comprising multiple interactions with different kinds of regulatory outcomes at both national and international levels with a range of outcomes.

Gulbrandsen (2014) identifies four ways in which private forms of environmental governance interact with and influence state regulation and policy. First, improvements in private standards can be realised when public procurement programmes are in place. By setting normative and cognitive demands for sustainable products, but outsourcing verification to private initiatives, states can set a benchmark with which private standards have to comply in order to compete (e.g. Ward and Phillips, 2008). Second, if states rely on private initiatives to implement regulation or set public policy goals the legitimacy of both states and private initiatives improves. If standards have industry support but are based on publicly set rules and expectations, a so-called race to the top is possible (Hilbrands, 2009). Third, states may elect to reassert authority by assessing credible standards when private standards emerge in densely regulated sectors such as seafood. States might seek such authority either bilaterally by setting their own standards or multilaterally through, for example, FAO codes of conduct. Fourth, a process of mutual learning may be possible when private initiatives disclose information about their impact or the outcomes of public regulation – for instance, if the organisations which own standards or traceability systems report their results and advocate for regulatory change (Auld et al., 2009).

Others have noted that states interact with private initiatives in more direct ways. As Foley (2013) notes, states facilitate commercial interest by funding the application of or conformity to private governance initiatives. States also play a more fundamental role in markets by taking on a dual position of regulator and producer (Foley, 2012; Adolf et al., 2015). By doing so, they directly intervene in struggles over surplus value and influence the environmental conditions of production (cf. Campling and Havice, 2014). It therefore appears essential to deepen our understanding of public–private governance interactions to include what Gale and Haward (2011) label 'hybrid' forms of governance that transcend outmoded public/private divisions.

Market-based seafood governance?

The increasingly blurred distinction of the public/private divide also raises questions about the extent to which 'private' sustainability governance is 'market based'. While market exchange, based on supply and demand, is commonly thought to drive the sustainable seafood movement, it is increasingly clear that there is a range of alternative logics driving private forms of governance.

For example, market-based arrangements governing access to fish stocks are ostensibly based on privatising rights to public resources with the intention of creating greater individual stewardship over the resource base (e.g. through ITQs) (Honneland, 2012). By capping total fishing effort, based on sound scientific advice, and allowing for the trade of ITQs, markets are expected to allocate fish to those who can most efficiently catch them. But as variously outlined (Allison, 2001; Degnbol et al., 2006; Mansfield, 2006, 2007), the universal application of individual rights as a solution to manage scarce fisheries' resources is highly problematic, because individual fishers embedded in their institutional, cultural and social context may not necessarily be the rational actors that economic theory expects them to be.

At a global scale, private mechanisms such as certification are based on a market logic: if consumers demand more sustainable production practices then producers have little choice but to respond. While this underlying assumption holds to a certain degree, it is not the only mechanism driving change. Private initiatives may work as much through a civic notion of moral responsibility that operates outside, or at the edges of, rational market demand (Ponte, 2012). Producers may wish, for example, to establish a social licence to operate with locally embedded actors as much as they want to sell sustainable products. States may also use private initiatives for their own goals, again outside any form of market logic. Finally, market actors such as retailers may use private governance arrangements to reduce reputational risk, which may not correspond to any specific demand for the products they sell (as also noted by Cashore, 2002). The consequence is that these so-called market-based mechanisms may be embedded and used by market actors, and even reinforce the relations of production and consumption that structure that market (Konefal, 2013). But their specific mode of operation may be far more dependent on broader regulatory networks (Vandergeest, 2007) that surround the market than on economic exchange alone.

Conclusions on governance

The globalised nature of production systems such as seafood has led to the emergence of new sustainability governance arrangements. The complexity of the global seafood provision system means it is unlikely that any single particular governance arrangement will prevail. Instead, governance entails the emergence and implementation of multiple initiatives and arrangements working across a range of scales from local to global, from sites of production to places of consumption and through a complex web of socio-spatial interactions. Unpacking these governance arrangements requires first opening up key challenges, such as between centralisation and decentralisation, between concentration and separation of power and spheres of control, and between public and private actor involvement. But while these challenges describe processes inherent to global seafood sustainability governance, they run up against explanatory limits because they ignore how social relations of production and consumption underscore processes of inclusion and exclusion, as well as how actors come together and organise themselves around the normative goal of sustainable seafood. Such questions are addressed through three related areas of scholarship: GVC analysis, the sociology of networks and flows, and social movements.

Governing chains and networks

Global value chains

GVC analysis has emerged in response to calls for a more comprehensive analysis of the organisation and control of transnational commodity flows (Gereffi, 1999, 2013; Gereffi et al., 2005; Gereffi and Lee, 2012). The central tenet of the GVC perspective is the identification of how power is distributed through different modes of coordination by so-called 'lead firms' over production and adding value (Gibbon and Ponte, 2005). By focusing on these firms, it is possible to identify not only the power relationships among suppliers but also the wider governance and market dynamics that emerge through and in response to global economic activity (Coe et al., 2008; Cheyns and Riisgaard, 2014).

A central preoccupation with GVC analysis is to categorise the types of coordination and control exerted by lead firms. Gereffi (1994) first distinguished buyer- and supplier-driven GVCs. Buyer-driven chains are dominated by retailing companies, such as in most food products, and supplier-driven chains by companies involved in primary production, such as in the automobile industry. This dichotomy has since been widely critiqued as being too simplistic, not taking into account the socially embedded character of transactions along the chain. Gereffi et al. (2005) responded by elaborating a set of five modes of GVC governance: market, modular, relational, captive and hierarchy. The classification of these modes of chain coordination is determined by the complexity of the transaction involved, the level of codification set over a product and the capability of suppliers to respond to these requirements. While providing an enduring understanding of the form and function of global chains, this classification has been critiqued as being too narrow because it assumes 'whole chain' coordination rather than multiple forms of coordination at different parts of the chain, and ignores the 'horizontal' or extra-transactional relations that shape economic activity within chains.

More recent approaches have remedied the preoccupation with the vertical dynamics of value chain governance by focusing attention on horizontal relations and dynamics. One of the first attempts to recognise the role of horizontal influences over chains was Gibbon et al.'s (2008) notion of governance as normalisation – referring to the how external non-firm actors, such as consumers, NGOs and government institutions, structure flows of information that are taken up in norms and practices by firm actors and in turn influence practices within and around a particular chain. More recently, Ponte and Sturgeon (2013) have introduced a modular approach to chain governance, surrounding Gereffi et al.'s (2005) 'meso-level' modes of coordination with micro-level social conventions and macro-level institutional analysis. But further work is needed. As Gereffi and Lee (2016) noted, the involvement of the state and NGOs in GVCs in terms of setting trade and resource regulation and standards, and training and support for upgrading, remains undertheorised. This, they argue, requires us to go beyond regulation alone to understand value chain governance as the synthesis of private (corporate), social (NGO) and public (state) influence over social and environmental outcomes.

More networked approaches better suited to the challenge of understanding global trade are in fact already well established. An alternative framework that captures these horizontal and vertical dimensions is the global production network (GPN) approach (Whatmore and Thorne, 1997; Coe et al., 2008; Hughes et al., 2008). GPN analysis looks explicitly at the 'globally organised nexus of interconnected functions and operations by firms and non-firm institutions through which goods and services are produced and distributed' (Coe et al., 2004: 471). The GPN approach thus focuses attention on non-firm actors in structuring global economic relations, and the regulatory and institutional contexts which shape production (Hess, 2004; Coe and Hess, 2005). As argued by Coe (2012), the emphasis on networks between and within 'territories' also underlines the spatial geographies of these economic relations in the context of globalisation, allowing for multiple ways of interaction between involved actors as opposed to the more structured typology of the GVC approach.

However, the GPN approach has its limitations. Despite focusing on networks, the vertical/linear movement of goods from producers to consumers remains central to most studies. Although other flows have been brought in (e.g. financial, informational, environmental), these are often limited to those directly of relevance to the bounded problematic of 'production-consumption'. In response, attention has been paid to nested processes or networks within GPNs. Examples include following financial flows through the locally embedded networks of cities (Sassen, 2010), and understanding the local social relations of production that facilitate the illegalities of ship breaking (Crang et al., 2013). However, GPN researchers have not extended their approach to sustainability and the environment beyond the relations of natural resource exploitation.

Despite their differences, the GVC and GPN approaches are increasingly grouped together as a single family (e.g. Derudder and Witlox, 2010; Dennis Wei et al., 2011). Indeed, in the introduction to their modular approach to GVC analysis, Ponte and Sturgeon (2013) go so far as to say that the differences between the frameworks is largely discursive, stressing their similar interests in mapping the spatial and organisational division of labour, the creation, capture and location of value, the institutional and regulatory processes, and the wider cultural economy influencing economic organisation. However, further innovation is needed for these schools of thought to help us understand more complex settings such as the sustainable seafood movement. One such innovation, we argue, is to link the challenges of governing global sustainable production and consumption to theories of networks and flows.

Networks and flows

The sociology of global environmental flows and networks (Spaargaren et al., 2006b; Oosterveer, 2009) acknowledges the importance of both material and non-material elements and the presence of multiple actors in sustainability governance (Guy and Marvin, 1996). According to this perspective, material flows analysis should be combined with the study of related immaterial flows, such as information and communication, and the relevant social actors that form part of these social networks. Contributions from different authors (Appadurai, 1996; Castells, 1996, 1997, 1998, 2009; Urry, 2000a, 2000b, 2003, 2007) can be drawn upon in further developing the analytical tools to study not just the material substances, economic dynamics and technical infrastructures in global seafood supply chains but also the societal networks, discourses and power relationships that go along with them.

Networks are dynamic structures, constraining and enabling agents (Giddens, 1984), because they steer roles and modes of operation with distinct sets of rules, while actors use networks as an enabling environment. Actors and practices in networks are connected across time and space through flows that include material products, monetary streams, knowledge and information, energy, etc. (Spaargaren et al., 2006a, 2006b). As argued by Ponte (2012), networks enable actors to deploy materials and techniques to enrol other actors, apply, adapt and 'translate' standards within localised spaces and extend their application beyond these spaces. In the context of seafood, this means bringing together variously defined global networks and flows that link humans, fish, ecosystems and other actors and intermediaries in multiple ways (Bear and Eden, 2008).

Adopting networks and flows as the main units of analysis allows a concentration on relational processes, i.e. interactions among social actors and between social actors and ecosystems. This perspective acknowledges the relevance of formal and informal arrangements organised by consumer/environmental associations, public sector representatives and actors within the relevant industry (Wilkinson, 2006). As Glin et al. (2012) argue, a networks and flows approach allows a more detailed analysis of the situated social and socio-material practices interacting both horizontally and vertically through multiple spaces, actors and networks. In the case of seafood, global networks are based on transnational trade relations. But in many respects seafood provision remains bounded to real-time and specific local space, or what Castells (1996) labels 'the space of place'. That is, where fish are caught or farmed remains dependent on the particular local ecological and aquatic conditions, and seafood consumption remains embedded in locally specific cultures and practices (Watson and Caldwell, 2005). The different dynamics at the local compared with the global level creates a tension that seafood governance has to address through network governance.

Networks become increasingly relevant in contemporary society because they are the primary drivers in the process of globalisation, facilitated by innovations in information and communications technology (ICT), the expansion of transnational corporations and the shifting role of nation-states (Castells, 2005). Or, as Castells (1996) argues, we are moving towards a 'global network society'. Increasingly, economic, social and political activities are organised through global networks, although not all contemporary networks are global. Networks include only those actors and practices that are deemed relevant from their internal perspective while ignoring other actors and practices. They may acquire different configurations through variations in the size and density of their networked connections internally as well as through their links with other networks. In the transition to global modernity, global networks and flows are increasingly replacing geographically bounded systems as they are embedded in multi-scalar regulatory contexts

(Coe et al., 2008). Network approaches to global sustainability governance are being introduced as 'regulatory networks' (Vandergeest, 2007), 'multi-level governance' (Mol, 2001) and 'hybrid transnational public-private regulatory systems' (Baird and Quastel, 2011), because it is no longer possible to separate states distinctly from markets or the public from the private sphere.

Human agency is essential when trying to understand how networks and flows are maintained and (re)constituted. Dynamics in networks result from the interactions among the human agents involved and their engagements with the relevant (im)material flows (Giddens, 1984; Elder-Vass, 2010). Although no individual human agent can fully control the complex dynamics of networks, no networks exist without human actors and their agency either. The differentiated capacity to achieve outcomes based on access to and use of material and discursive resources (Giddens, 1984) is relevant for understanding this dynamic. The power of a particular social actor in a network is determined by the position (s)he occupies within that network and the differences in access to and control over resources. Most actors have a generic position within global networks without much opportunity to influence the functioning of these networks, nor their roles in these networks. However, it is clear that some actors have more possibilities to steer networks in the contemporary global network society than others.

Beyond the conventional economic and political sources of power, Castells (2009) identifies two additional sources: *programming* and *switching* (i.e. connecting) networks. These new forms of power are based on discursive rather than material or political resources. Programmers determine the crucial mechanisms for inclusion and exclusion in a network (Cheyns and Riisgaard, 2014) and its mode of operation, including 'ideas, visions, projects and frames' (Castells, 2009). By combining information and knowledge in defining the network goals and how to achieve them, programmers (re)constitute the network itself and establish a way of operating smoothly. The other category of emerging power holders in the network society, the switchers, determines how one network engages with other networks and controls the access from one network to another. Switchers steer the connections between different networks by defining common goals and combining resources to strengthen their position within a 'network of networks', and thereby also their operational effectiveness. These connections are constructed at specific interfaces and are generally stable (ibid.). Programming and switching are not necessarily identified with one particular individual agent, but often with groups at the interface between various social networks (Castells, 2011). Although less visible than the conventional holders of economic and political power, programmers and switchers nevertheless have influential positions in the contemporary global network society.

The concept of global networks and flows offers a promising starting point for analysing sustainability governance in global seafood provision. It includes the material as well as the social dimensions, acknowledges complexity, dynamics and uncertainty and looks beyond the internal economic dynamics of global supply chains or the political dynamics of nation-states and multilateral agreements. A networks and flows perspective instead looks at how global seafood provision is embedded in local, regional and global socio-material relations. But recognising the multiple levels involved in global seafood provision, this perspective does not privilege any one scale, because the constitution of these networks and flows cannot be known beforehand. The global and local dynamics in seafood provision are inextricably and irreversibly bound together through dynamic relationships within networks and flows.

Social movements

Social movements are highly relevant to contemporary seafood governance because they put pressure on governments that fail to act and induce changes in the practices of private companies and consumers (Prakash and Potoski, 2007). In general, social movements can be defined as 'an association or set of associations organised around a common interest that seeks to influence collective outcomes without obtaining authoritative offices of government' (Dryzeck et al., 2003: 2). This very general definition covers the many different initiatives that exist: the degree of formal institutionalisation; the scale or level at which they operate; and the mode of direct-action lobbying, public education and awareness raising they employ (Diani and Donati, 1999). Social movements may emerge at local (Diani and Rambaldo, 2007; Rootes, 2007), national (Jamison et al., 1990) or global scales (Anheier et al., 2002), but common to all scales are networks of social relations. How these social relations constitute and are constitutive of networks and flows is a promising starting point for understanding how they develop and what effects they can have in attaining their normative and cognitive goals.

Social movements are conceptualised in terms of 'resource mobilisation' or 'identity building' (Hannigan, 1985; Benford and Snow, 2000). The resource-mobilisation perspective argues that 'contemporary social movements are composed of a plurality of organisations and groups engaged in strategic action on a political field, competing and bargaining with their counterparts from the established political culture, as well as with each other' (Jamison et al., 1990: 3). Social movements, according to this view, deploy different (material, human, cognitive, technical and organisational) resources to build their organisation and achieve the goals they have set themselves (Edelman, 2001). Social movements are essentially networks of interest groups operating in a rational, strategic manner whereby the possibilities of achieving their goals relate to the 'political opportunity structure' provided by the state and their capacity to mobilise. From the 'identity-building' perspective, contemporary social movements are attempts to form new collective identities in late-modern societies (Jamison et al., 1990). These new social movements therefore emphasise 'quality of life and lifestyle concerns' rather than 'economic redistribution' (Pichardo, 1997: 414).

Returning to Castells, the new social movements are a central component of globalisation, as they seek to produce new values and goals around which institutions of society are transformed to represent these values by 'creating new norms to organise social life' (Castells, 2012: 9). They do this as they move beyond the space of flows to the space of place, directly engaging actors in everyday practices of consumption and production.

Since their start in the 1970s, environmental social movements in general have tried to steer politics into a more environment-friendly direction, and have become significant in the social shaping of science and technology (Jamison, 1996) and more recently in influencing the market. The justification for this 'market turn' is, as outlined under the governance section above, seen as a collective frustration with the limited impacts of environmental regulation by the state. In the sustainable seafood movement, as in many other environmental social movements, the emphasis has shifted away from traditional modes of organisation and advocacy to new patterns of resistance and regulation that seek to manage production practices without directly involving public authorities. The movement has shifted from unilateral modes of NGO action to a range of so-called sustainability partnerships related to both deregulation and internationalisation of markets and products that have persistent and complex sustainability issues (Glasbergen, 2007; Bitzer et al., 2012). These partnerships aim to create collaborative institutionalised arrangements between actors from two or more sectors of society – market, state and civil society – aimed at the provision and protection of collective goods through GVCs (Bitzer and Glasbergen, 2015). Glasbergen (2007) postulates that the degree of influence states have in these partnerships depends on the coordinating power of the commercial partner in the value chain. As such, the conditions of these partnerships are shaped by political and economic competition, conflict and cooperation between firm and non-firm actors both in and around these chains.

But concerns have been raised about the diminishing accountability of civil society as these partnerships became professionalised, to the extent that NGOs speak for firms rather than challenge them to change (e.g. Baur and Schmitz, 2012; Robinson, 2012). Partnerships represent a careful balance in engaging a more extensive range of actors with often very different values and goals. They can also lead to creative innovation and disruptive change by offering new rationalities around sustainability in processes of both production and trade, and bring them to the same level as more traditional economic rationalities (Oosterveer and Spaargaren, 2011). Success is then the extent that environmental considerations are taken up in the policy and practice of firms and governments. From a more critical perspective, however, the potential for these partnerships to create fundamental change is questioned, because the approaches that underlie these partnerships seek to build on rather than change the capitalist modes of production that lead to environmental degradation in the first place (Konefal, 2013). As civil society actors are drawn further into these partnerships and become more reliant on the firms and states they seek to change, their effectiveness may be undermined.

Given the different possible trajectories and approaches that constitute environmental movements, it is necessary to rethink how they can create change towards more sustainable production and consumption. This is all the more necessary when we are trying to understand how these movements operate in the context of a globalised food system such as seafood.

Towards a new understanding of the sustainable seafood movement

Globalisation in seafood provision coincides with increased (awareness of) sustainability problems, and dealing with them means facing complex challenges. Among the complications are how to understand these changes in seafood provision, how to balance the roles of public versus private actors and how to develop inclusive governance arrangements. The complexity of present-day seafood provisioning networks necessitates a more interactive/reflective mode of governance (Meadowcroft, 2007) and leads to a number of relevant questions. If the seafood movement is a mosaic, how can we understand the emergence of innovative governance mechanisms? How do these operate in chains and networks? To what extent can this movement foster change? How can we understand the seafood movement? What forms of governance are likely to be more inclusive and foster improvement, and who do they involve?

Scholars are struggling with questions over how we can cope with the shift away from government, including the legitimacy and authority of non-state actors, how state actors are involved in hybrid arrangements and the extent and effectiveness of so-called market-based approaches. All these questions point to where and how the shift to new forms of sustainability governance can influence processes of global seafood production and consumption. In other words, the challenge is to determine who can govern on whose behalf on a global scale once we move beyond state-based territorial dimensions of control (Sassen, 2006). We argue that by combining value chains, networks and flows and social movements we can identify the dynamic way in which the sustainable seafood movement is put together over global space. This combination has three potential advantages. Value chains highlight the vertical role of lead firms and the dynamic nature of control over material and informational flows within and around these chains. The theory of global networks and flows can be applied as a conceptual framework to analyse global sustainable seafood governance by combining local-level and global-level dynamics while linking social and material dimensions. Finally, social movements point to the changing forms of identity and symbolism in shifting everyday practices of consumption towards sustainability. It also helps to question what modes of collective agency, represented by civil society organisations such as NGOs, can lead to effective modes of change and transition on a global scale.

By employing these approaches in our analysis of the sustainable seafood movement, we contribute to the selected governance themes we outlined at the start of this chapter. In short, we question how the movement can create legitimate forms of sustainability governance that are inclusive of a wide range of producers and buyers while steering long-term change in both private and public spheres of authority. The following chapter starts us on this journey by describing the global geographies of seafood production.

References

Adolf, S., S. R. Bush and S. Vellema (2015). 'Reinserting state agency in global value chains: The case of MSC certified skipjack tuna', *Fisheries Research* 182, 79–87.

Allison, E. H. (2001). 'Big laws, small catches: Global ocean governance and the fisheries crisis', *Journal of International Development* 13(7), 933–950.

Anheier, H., M. Glasius and M. Kaldor (eds) (2002). *Global Civil Society Handbook, 2002*. Oxford: Oxford University Press.

Appadurai, A. (1996). *Modernity at Large. Cultural Dimensions of Globalization*. Minneapolis and London: University of Minnesota Press.

Auld, G., C. Balboa, S. Bernstein and B. Cashore (2009). 'The emergence of non-state market-driven (NSDM) global environmental governance: A cross-sectoral assessment', in M. A. Delmas and O. R. Young (eds) *Governance for the Environment. New Perspectives*. Cambridge: Cambridge University Press, 183–218.

Baird, I. G. and N. Quastel (2011). 'Dolphin-safe tuna from California to Thailand: Localisms in environmental certification of global commodity networks', *Annals of the Association of American Geographers* 101(2), 337–355.

Barkin, J. S. and E. R. DeSombre (2013). *Saving Global Fisheries: Reducing Fishing Capacity to Promote Sustainability*. Cambridge, MA: MIT Press.

Baur, D. and H. P. Schmitz (2012). 'Corporations and NGOs: When accountability leads to co-optation', *Journal of Business Ethics* 106(1), 9–21.

Bear, C. and S. Eden (2008). 'Making space for fish: The regional, network and fluid spaces of fisheries certification', *Social & Cultural Geography* 9(5), 487–504.

Benford, R. D. and D. A. Snow (2000). 'Framing processes and social movements: An overview and assessment', *Annual Review of Sociology* 26, 611–639.

Bitzer, V. and P. Glasbergen (2015). 'Business–NGO partnerships in global value chains: Part of the solution or part of the problem of sustainable change?', *Current Opinion in Environmental Sustainability* 12, 35–40.

Bitzer, V., P. Glasbergen and P. Leroy (2012). 'Partnerships of a feather flock together? An analysis of the emergence of networks of partnerships in the global cocoa sector', *Global Networks* 12(3), 355–374.

Boström, M. (2006). 'Regulatory credibility and authority through inclusiveness: Standardization organizations in cases of eco-labelling', *Organization* 13(3), 345–367.

Boström, M. and M. Klintman (2008). *Eco-standards, Product Labelling and Green Consumerism*. Houndmills: Palgrave Macmillan.

Campbell, H. (2005). 'The rise and rise of EurepGAP: European (re)invention of colonial food relations?', *International Journal of Food and Agriculture* 13(2), 1–19.

Campling, L. and E. Havice (2014). 'The problem of property in industrial fisheries', *Journal of Peasant Studies* 41(5), 707–727.

Casey, D. (2008). 'Three puzzles of private governance: GlobalGAP and the regulation of food safety and quality', in *(Re)Regulation in the Wake of Neoliberalism. Consequences of Three Decades of Privatization and Market Liberalization*. Utrecht: ECPR Standing Group on Regulatory Governance, 26.

Cashore, B. (2002). 'Legitimacy and the privatization of environmental governance: How non-state market-driven (NSMD) governance systems gain rule-making authority', *Governance* 15(4), 503–529.

Cashore, B., G. Auld and D. Newsom (2004). *Governing Through Markets: Forest Certification and the Emergence of Non-State Authority*. New Haven, CT: Yale University Press.

Cashore, B., G. Auld, S. Bernstein and C. McDermott (2007). 'Can non-state governance "ratchet up" global environmental standards? Lessons from the forest sector', *Review of European Community & International Environmental Law* 16(2), 158–172.

Castells, M. (1996). *The Rise of the Network Society. Volume I of The Information Age: Economy, Society and Culture*. Malden, MA and Oxford: Blackwell.

Castells, M. (1997). *The Power of Identity. Volume II of The Information Age: Economy, Society and Culture*. Malden, MA and Oxford: Blackwell.

Castells, M. (1998). *End of Millenium. Volume III of The Information Age: Economy, Society and Culture*. Malden, MA and Oxford: Blackwell.

Castells, M. (2005). 'The network society: From knowledge to policy', in M. Castells and G. Cardoso (eds) *The Network Society: From Knowledge to Policy*. Washington, DC: Center for Transatlantic Relations, 3–21.

Castells, M. (2009). *Communication Power*. Oxford: Oxford University Press.

Castells, M. (2011). 'A network theory of power', *International Journal of Communication* 5, 773–787.

Castells, M. (2012). *Networks of Outrage and Hope: Social Movements in the Internet Age*. Cambridge: Polity Press.

Cheyns, E. (2011). 'Multi-stakeholder initiatives for sustainable agriculture: Limits of the "inclusiveness" paradigm', in S. Ponte, P. Gibbon and J. Vestergaard (eds) *Governing through Standards: Origins, Drivers and Limitations*. Houndmills, Palgrave Macmillan, 210–235.

Cheyns, E. and L. Riisgaard (2014). 'Introduction to the symposium', *Agriculture and Human Values* 31(3), 409–423.

Chuenpagdee, R. and S. Jentoft (2009). 'Governability assessment for fisheries and coastal systems: A reality check', *Human Ecology* 37(1), 109–120.

Coe, N. M. (2012). 'Geographies of production II: A global production network A–Z', *Progress in Human Geography* 36(3), 389–402.

Coe, N. M. and M. Hess (2005). 'The internationalization of retailing: Implications for supply network restructuring in East Asia and Eastern Europe', *Journal of Economic Geography* 5(4), 449–473.

Coe, N. M., P. Dicken and M. Hess (2008). 'Global production networks: Realizing the potential', *Journal of Economic Geography* 8(3), 271–295.

Coe, N. M., M. Hess, H. W.-C. Yeung, P. Dicken and J. Henderson (2004). '"Globalizing" regional development: A global production networks perspective', *Transactions of the Institute of British Geographers* 29(4), 468–484.

Constance, D. H. and A. Bonanno (1999). 'Contested terrain of the global fisheries: "Dolphin-safe" tuna, the Panama Declaration, and the Marine Stewardship Council', *Rural Sociology* 64(4), 597–623.

Constance, D. H. and A. Bonanno (2000). 'Regulating the global fisheries: The World Wildlife Fund, Unilever and the Marine Stewardship Council', *Agriculture and Human Values* 17, 171–187.

Constanza, R., F. Andrade, P. Antunes, M. van den Belt, D. Boesch, D. Boersma, F. Catarino, S. Hanna, K. Limburg, B. Low and M. Molitor (1999). 'Commentary:

Ecological economics and sustainable governance of the oceans', *Ecological Economics* 31, 171–187.

Crang, M., A. Hughes, N. Gregson, L. Norris and F. Ahamed (2013). 'Rethinking governance and value in commodity chains through global recycling networks', *Transactions of the Institute of British Geographers* 38(1), 12–24.

Degnbol, P., H. Gislason, S. Hanna, S. Jentoft, J. R. Nielsen, S. Sverdrup-Jensen and D. C. Wilson (2006). 'Painting the floor with a hammer: Technical fixes in fisheries management', *Marine Policy* 30(5), 534–543.

Dennis Wei, Y., I. Liefner and C.-H. Miao (2011). 'Network configurations and R&D activities of the ICT industry in Suzhou municipality, China', *Geoforum* 42(4), 484–495.

Derudder, B. E. N. and F. Witlox (2010). 'World cities and global commodity chains: An introduction', *Global Networks* 10(1), 1–11.

DeSombre, E. R. and J. S. Barkin (2011). *Fish*. Cambridge: Polity Press.

Diani, M. and P. R. Donati (1999). 'Organisational change in Western European environmental groups: A framework for analysis', *Environmental Politics* 8(1), 13–34.

Diani, M. and E. Rambaldo (2007). 'Still the time of environmental movements? A local perspective', *Environmental Politics* 16(5), 765–784.

Dryzeck, J. S., D. Downes, C. Hunold and D. Schlosberg (2003). *Green States and Social Movements: Environmentalism in the United States, United Kingdom, Germany, and Norway*. Oxford and New York: Oxford University Press.

Edelman, M. (2001). 'Social movements: Changing paradigms and forms of politics', *Annual Review of Anthropology* 30, 285–317.

Elder-Vass, D. (2010). *The Causal Power of Social Structures. Emergence, Structure and Agency*. Cambridge: Cambridge University Press.

Eshuis, J. and A. Edwards (2013). 'Branding the city: The democratic legitimacy of a new mode of governance', *Urban Studies* 50(5), 1066–1082.

Finley, C. (2011). *All the Fish in the Sea: Maximum Sustainable Yield and the Failure of Fisheries Management*. Chicago, IL: University of Chicago Press.

Foley, P. (2012). 'The political economy of Marine Stewardship Council certification: Processors and access in Newfoundland and Labrador's inshore shrimp industry', *Journal of Agrarian Change* 12(2/3), 436–457.

Foley, P. (2013). 'National government responses to Marine Stewardship Council (MSC) fisheries certification: Insights from Atlantic Canada', *New Political Economy* 18(2), 284–307.

Fuchs, D., A. Kalfagianni and T. Havinga (2011). 'Actors in private food governance: The legitimacy of retail standards and multistakeholder initiatives with civil society participation', *Agriculture and Human Values* 28(3), 353–367.

Fulponi, L. (2006). 'Private voluntary standards in the food system: The perspective of major food retailers in OECD countries', *Food Policy* 31, 1–13.

Gale, F. and M. Haward (2011). *Global Commodity Governance: State Responses to Sustainable Forest and Fisheries Certification*. London: Palgrave Macmillan, 173–206.

Gereffi, G. (1994). 'The organization of buyer-driven global commodity chains: How U.S. retailers shape overseas production networks', in G. Gereffi and M. Korzeniewicz (eds) *Commodity Chains and Global Capitalism*. Westport, CT: Praeger, 95–122.

Gereffi, G. (1999). 'International trade and industrial upgrading in the apparel commodity chain', *Journal of International Economics* 48, 37–70.

Gereffi, G. (2013). 'Global value chains in a post-Washington Consensus world', *Review of International Political Economy* 21(1), 1–29.

Gereffi, G. and J. Lee (2012). 'Why the world suddenly cares about global supply chains', *Journal of Supply Chain Management* 48(3), 24–32.

Gereffi, G. and J. Lee (2016). 'Economic and social upgrading in global value chains and industrial clusters: Why governance matters', *Journal of Business Ethics* 133(1), 25–38.

Gereffi, G., J. Humphrey and T. Sturgeon (2005). 'The governance of global value chains', *Review of International Political Economy* 12(1), 78–104.

Gibbon, P. and S. Ponte (2005). *Trading Down. Africa, Value Chains, and the Global Economy*. Philadelphia, PA: Temple University Press.

Gibbon, P., J. Bair and S. Ponte (2008). 'Governing global value chains: An introduction', *Economy and Society* 37(3), 315–338.

Giddens, A. (1984). *The Constitution of Society. Outline of the Theory of Structuration*. Cambridge: Polity Press.

Glasbergen, P. (2007). 'Setting the scene: The partnership paradigm in the making', in P. Glasbergen, F. Biermann and A. P. J. Mol (eds) *Partnerships, Governance and Sustainable Development: Reflections on Theory and Practice*. Cheltenham: Edward Elgar, 1–28.

Glin, L. C., P. Oosterveer, A. P. J. Mol and S. D. Vodouhe (2012). 'Governing the transnational organic cotton network from Benin', *Global Networks* 12, 333–354.

Gulbrandsen, L. H. (2012). *Transnational Environmental Governance: The Emergence and Effects of the Certification of Forests and Fisheries*. Cheltenham: Edward Elgar.

Gulbrandsen, L. H. (2014). Dynamic governance interactions: Evolutionary effects of state responses to non-state certification programs', *Regulation & Governance* 8(1), 74–92.

Guy, S. and M. J. Marvin (1996). 'Transforming urban infrastructure provision – The emerging logic of demand side management', *Policy Studies* 17(2), 137–147.

Hannigan, J. A. (1985). 'Alain Touraine, Manuel Castells and social movement theory: A critical appraisal', *Sociological Quarterly* 26(4), 435–454.

Hatanaka, M., C. Bain and L. Busch (2005). 'Third-party certification in the global agrifood system', *Food Policy* 30, 354–369.

Held, D., A. McGrew, D. Goldblatt and J. Perraton (1999). *Global Transformations. Politics, Economics and Culture*. Cambridge: Polity Press.

Hess, M. (2004). '"Spatial" relationships? Towards a reconceptualization of embedded ness', *Progress in Human Geography* 28(2), 165–186.

Hilbrands, A. (2009). 'GlobalGAP – Building consumer confidence in aquaculture sustainability', in T. Ward and B. Phillips (eds) *Seafood Ecolabelling: Principles and Practice*. Oxford: Wiley-Blackwell, 134–142.

Hoel, A. H., A. K. Sydnes and S. A. Ebbin (2005). 'Ocean governance and institutional change', in S. A. Ebbin, A. H. Hoel and A. K. Sydnes (eds) *A Sea Change: The Exclusive Economic Zone and Governance Institutions for Living Marine Resources*. Dordrecht: Springer, 3–16.

Honneland, G. (2012). *Making Fishery Agreements Work. Post-Agreement Bargaining in the Barents Sea*. Cheltenham and Northampton: Edward Elgar.

Hughes, A., N. Wrigley and M. Buttle (2008). 'Global production networks, ethical campaigning, and the embeddedness of responsible governance', *Journal of Economic Geography* 8, 345–367.

Iles, A. (2007). 'Making the seafood industry more sustainable: Creating production chain transparency and accountability', *Journal of Cleaner Production* 15(6), 577–589.

Imperial, M. T. and T. Yandle (2005). 'Taking institutions seriously: Using the IAD framework to analyze fisheries policy', *Society & Natural Resources* 18(6), 493–509.

Jacquet, J. L. and D. Pauly (2007). 'The rise of seafood awareness campaigns in an era of collapsing fisheries', *Marine Policy* 31(3), 308–313.

Jamison, A. (1996). 'The shaping of the global envionmental agenda: the role of non-governmental organisations', in S. Lash, B. Szerszynski and B. Wynne (eds) *Risk, Environment and Modernity. Towards a New Ecology*. London: Sage, 224–245.

Jamison, A., R. Eyerman, J. Cramer and J. Laessoe (1990). *The Making of the New Environmental Consciousness. A Comparative Study of the Environmental Movements in Sweden, Denmark and the Netherlands*. Edinburgh: Edinburgh University Press.

Jentoft, S. (2000). 'The community: A missing link of fisheries management', *Marine Policy* 24(1), 53–60.

Juda, L. (2002). 'Rio Plus Ten: The evolution of international marine fisheries governance', *Ocean Development & International Law* 33, 109–144.

Kelleher, K., R. Willmann and R. Arnason (2009). *The Sunken Billions: The Economic Justification for Fisheries Reform*. Washington, DC: World Bank.

Konefal, J. (2013). 'Environmental movements, market-based approaches, and neoliberalization: A case study of the sustainable seafood movement', *Organization & Environment* 26(3), 336–352.

Konefal, J., M. Mascarenhas and M. Hatanaka (2005). 'Governance in the global agro-food system: Backlighting the role of transnational supermarket chains', *Agriculture and Human Values* 22(3), 291–302.

Kooiman, J. and M. Bavinck (2005). 'The governance perspective', J. Kooiman, S. Jentoft, M. Bavinck and R. Pullin (eds) *Fish for Life: Interactive Governance for Fisheries*. Amsterdam: Amsterdam University Press, 11–24.

Mansfield, B. (2006). 'Assessing market-based environmental policy using a case study of North Pacific fisheries', *Global Environmental Change* 16(1), 29–39.

Mansfield, B. (2007). 'Articulation between neoliberal and state-oriented environmental regulation: Fisheries privatization and endangered species protection', *Environment and Planning A* 39(8), 1926–1942.

Marsden, T., R. Lee, A. Flynn and S. Thankappan (2010). *The New Regulation and Governance of Food: Beyond the Food Crisis?* New York and London: Routledge.

Meadowcroft, J. (2007). 'Who is in charge here? Governance for sustainable development in a complex world', *Journal of Environmental Policy & Planning* 9(3–4): 299–314.

Miller, A. M. and S. R. Bush (2015). 'Authority without credibility? Competition and conflict between ecolabels in tuna fisheries', *Journal of Cleaner Production* 107, 137–145.

Mohammed, E. Y. (2014). 'Fisheries and the post-2015 development agenda', IIED briefing paper. London: IIED.

Mol, A. P. J. (2001). *Globalization and Environmental Reform. The Ecological Modernization of the Global Economy*. Cambridge, MA: MIT Press.

Nadvi, K. and F. Wältring (2004). 'Making sense of global standards', in Hubert Schmitz (ed.) *Local Enterprises in the Economy: Issues of Governance and Upgrading*. Cheltenham: Edward Elgar, 53–94.

Oosterveer, P. J. M. (2009). 'Governing global environmental flows: Ecological modernization in technonatural time/spaces', in D. F. White and C. Wilbert (eds) *Technonatures: Environments, Technologies, Spaces, and Places in the Twenty-first Century*. Waterloo, ON: Wilfrid Laurier University Press, 35–62.

Oosterveer, P. and D. A. Sonnenfeld (2012). *Food, Globalization and Sustainability*. London and New York: Earthscan.

Oosterveer, P. and G. Spaargaren (2011). 'Organising consumer involvement in the greening of global food flows: The role of environmental NGOs in the case of marine fish', *Environmental Politics* 20(1), 97–114.

Pattberg, P. H. (2007). *Private Institutions and Global Governance: The New Politics of Environmental Sustainability*. Cheltenham: Edward Elgar.

Pichardo, N. A. (1997). 'New social movements: A critical review', *Annual Review of Sociology* 23, 411–430.

Ponte, S. (2012). 'The Marine Stewardship Council (MSC) and the making of a market for "sustainable fish"', *Journal of Agrarian Change* 12(2–3), 300–315.

Ponte, S. and T. Sturgeon (2013). 'Explaining governance in global value chains: A modular theory-building effort', *Review of International Political Economy* 21(1), 1–29.

Prakash, A. and M. Potoski (2007). 'Collective action through voluntary environmental programs: A club theory perspective', *Policy Studies Journal* 35(4), 773–792.

Rantala, T. (2012). 'Legitimacy of forest and nature conservation policy: A conceptual framework with illustrations', *Scandinavian Journal of Forest Research* 27(2), 164–176.

Robinson, J. G. (2012). 'Common and conflicting interests in the engagements between conservation organizations and corporations', *Conservation Biology* 26(6), 967–977.

Rootes, C. (2007). Acting locally: The character, contexts and significance of local environmental mobilisations. *Environmental Politics* 16(5), 722–741.

Sala, E., J. Mayorga, C. Costello, D. Kroodsma, M. L. Palomares, D. Pauly, U. R. Sumaila and D. Zeller (2018). 'The economics of fishing the high seas', *Science Advances* 4(6), eaat2504.

Sassen, S. (2006). *Territory, Authority, Rights: From Medieval to Global Assemblages*. Princeton, NJ and Oxford: Princeton University Press.

Sassen, S. (2010). 'Global inter-city networks and commodity chains: Any intersections?', *Global Networks* 10(1), 150–163.

Sen, S. and J. Raakjaer Nielsen (1996). 'Fisheries co-management: A comparative analysis', *Marine Policy* 20(5), 405–418.

Spaargaren, G., A. P. J. Mol and H. Bruyninckx (2006a). 'Introduction: Governing environmental flows in global modernity', in G. Spaargaren, A. P. J. Mol and F. H. Buttel (eds) *Governing Environmental Flows. Global Challenges to Social Theory*. Cambridge, MA and London: MIT Press, 1–36.

Spaargaren, G., A. P. J. Mol and F. H. Buttel (eds) (2006b). *Governing Environmental Flows. Global Challenges to Social Theory*. Cambridge, MA and London: MIT Press.

Urry, J. (2000a). 'Mobile sociology', *British Journal of Sociology* 51(1), 185–203.

Urry, J. (2000b). *Sociology beyond Society*. London: Routledge.

Urry, J. (2003). *Global Complexity*. Cambridge: Polity Press.

Urry, J. (2007). *Mobilities*. Cambridge: Polity Press.

Van Riel, M. C., S. R. Bush, P. A. M. van Zwieten and A. P. J. Mol (2015). 'Understanding fisheries credit systems: Potentials and pitfalls of managing catch efficiency', *Fish and Fisheries* 16(3), 453–470.

Vandergeest, P. (2007). 'Certification and communities: Alternatives for regulating the environmental and social impacts of shrimp farming', *World Development* 35(7), 1152–1171.

Vandergeest, P. and A. Unno (2012). 'A new extraterritoriality? Aquaculture certification, sovereignty, and empire', *Political Geography* 31(6), 358–367.

Ward, T. J. and B. Phillips (2008). 'Anecdotes and lessons of a decade', in T. J. Ward and B. Phillips (eds) *Seafood Ecolabelling Principles and Practice*. Oxford: Blackwell, 416–435.

Watson, J. I. and M. I. Caldwell (eds) (2005). *The Cultural Politics of Food and Eating: A Reader*, Blackwell Readers in Anthropology. Malden, MA and Oxford: Blackwell.

Whatmore, S. and L. Thorne (1997). Nourishing networks. Alternative geographies of food. *Globalizing Food. Agrarian Questions and Global Restructuring*. D. Goodman, and Michael Watts. London, Routledge: 287–304.

Wilkinson, J. (2006). 'Network theories and political economy: From attrition to convergence?', in T. Marsden and J. Murdoch (eds) *Between the Local and the Global: Confronting Complexity in the Contemporary Agri-Food Sector*. Oxford: Elsevier, 11–38.

3 The global dynamics of sustainable seafood

Introduction

Seafood is the most globally produced and traded food commodity in the world. Yet just because seafood is a globalised commodity does not mean its production, trade and consumption are uniform around the globe. Underlying the global statistics of the FAO are a series of more nuanced patterns and geographies, which together explain how production is distributed across the globe, who benefits regarding income, trade and employment, and who benefits in terms of consumption. We argue that the apparent shifts in these seafood geographies are key to addressing the challenges faced by a 'global' Anglo-American sustainable seafood movement which aims to shape diverse local practices towards the achievement of sustainability goals.

We address these geographies in this chapter in two parts. First, we provide an overview of the global geographies of production and consumption. Second, we identify and review what we label the four 'social geographies' of global seafood that are relevant for understanding the current make-up of the sustainable seafood movement and its future development.

Geographies of capture fisheries' production

The common perception, backed up by global FAO statistics, is that capture fisheries' production has stagnated since around the mid-1980s. Before this time the expansion of marine capture, in particular, was spectacular. Driving this expansion were considerable investments in fishing capacity by industrialised nations, particularly countries of Europe, East Asia and North America. The development of these fisheries is modelled by the Sea Around Us project (e.g. Watson et al., 2004, 2013), showing the steady diffusion of fishing effort away from the North Sea and east coast of the US and Canada to West Africa and Latin America. Subsequently there was a boom in capacity in East Asia, driven mainly by Japanese and Taiwanese expansion into Southeast Asia and the Western Pacific tuna fisheries.

This expansion of production results from the increase of fishing effort in the world's oceans. As shown by Watson et al. (2013), the intensity of fishing effort for all countries has grown tenfold on average since FAO records begun in 1950. Technological advancements are responsible for this increase – driven in part by post-war and Cold War investment in sonar and GPS (global positioning systems; Anticamara et al., 2011), but also in response to changing access to the world's oceans and seas. The geographic pattern of this spatial expansion was predominantly driven by Asian and European fleets, both of which took advantage of the *mare liberum*, or so-called open access, nature of marine capture fisheries before the introduction of EEZs in 1982 (ibid.).

Once EEZs were in place, giving sovereignty to coastal states over 200 nautical miles of adjacent waters, these fishing nations adopted a range of alternative access arrangements. Gaining greater control over their waters was a seemingly important step to the enclosure of resources under national management regimes, because the majority of fisheries' production remained within EEZs. However, despite this enclosure fishing effort continued to expand in two main directions. First, coastal states invested in domestic fleets, increasing capacity through subsidies to industrial fishing vessels and small-scale coastal fisheries under the auspices of economic development and food security (Berkes, 1985). Second, foreign countries which had established fleets in the EEZs of other, predominantly developing, countries had to apply for licences to operate within these new national jurisdictions.

Despite the promise of greater control over fish stocks within EEZs, many developing countries, especially in fish-abundant regions such as Southeast Asia and West Africa, did not have the resources to exploit their coastal fisheries. Investment came from the industrialised fishing nations, which strived to reflag their vessels under these coastal states, extend financing through development assistance and the promotion of joint ventures, or create licensing arrangements that allowed long-term access to these coastal waters (Munro, 2004, 2007). As documented by Butcher (2004), Japanese investment drove much of the expansion in tuna and demersal fisheries in Indonesia and other areas of Southeast Asia through a variety of joint ventures with Indonesian processing companies. In countries such as Thailand, similar joint ventures were taken over by the government through state subsidies. Similarly, in Vietnam state involvement in fisheries led to expanded effort (Pomeroy et al., 2009). The result has been the most substantial continental expansion in fishing effort, amounting to a 25-fold increase since 1950 (Watson et al., 2013). In West Africa the EU has created agreements with coastal states since the 1970s; currently the EU has 12 Sustainable Seafood Partnership Agreements.[1] These were for a long time designed only from an economic interest perspective, and their primary goal was to access foreign fishery resources and secure supply to and therefore employment for European seafood processing industries (Kaczynski and Fluharty, 2002). Economic development of the partner countries was not part of the deal, and the UN Development Programme observed that 'after 15 years of "cooperation" with the European Union, Senegal's fishing sector is in deep crisis' (UNDP,

2005: 145). These are agreements between unequal partners and have negative social and environmental impacts (Bretherton and Vogler, 2008). Resource conservation and environmental sustainability are now being included in these arrangements through EU Sustainable Fisheries Partnership Agreements; but developing local fish-processing industries and contributing to promoting social and economic development are not.

The enclosure of the EEZs also pushed industrialised fleets to areas beyond national jurisdiction (or the high seas) as the next 'frontier' in global fisheries' production. This movement required increased investment and state subsidies, given the substantially lower productivity of these areas and the difficulty in finding and targeting fish. The fleets exploiting these high-seas areas were predominantly from the industrialised economies of Europe, North America and East Asia. However, other countries are increasingly invested in their high seas distant-water fleets, including China and Indonesia. Complicating the pattern of investment is a series of access agreements distant-water fishing nations have made with so-called 'flag states', comprising joint licensing arrangements with industrialised states. This has led to some perverse examples whereby some flag states have many vessels but no fleet of their own. The complexities in the flagging of vessels has led to a lack of transparency in these fisheries, which has contributed to illegal and unreported fishing – the cost of which is estimated at US$10–23.5 billion annually, representing 11–26 million MT of seafood per year (Agnew et al., 2009), while also increasing uncertainties around the health of fish stocks.

The expansion of fishing effort has contributed to the stagnation in the volume of fish landed in two ways. First, representing a more pervasive form of overfishing, fishing fleets in nearly all parts of the world's oceans need to work ever more intensely to land the same amount of fish. This means that catches have to be maintained by expanding fishing into new waters and increasing the efficiency of fishing technologies in existing fishing grounds. Second, next to this creeping increase in fishing effort there have been a series of fisheries' collapses. The early collapses were limited to what biologists have labelled 'R-selected' species, which are highly fecund and experience high fluctuations in reproduction levels. They include crises with species such as Peruvian anchovy and stocks of herring in the North Atlantic and the North Sea in the 1970s. However, as argued by Pauly (2008), these collapses did not raise much alarm because they were limited to specific pelagic stocks and were put down to environmental factors. It was only when a series of other significant stocks, such as North American cod, collapsed in the 1980s and 1990s that the pervasive nature of increased fishing effort, coupled with the substantial uncertainty in the status of these stocks (Finlayson, 1994), led to greater societal awareness about the consequences of continuing to expand or even maintain existing levels of fishing effort.

Today it is recognised that the systemic overcapacity of fishing fleets in the world's fisheries has had a tremendous impact on the ecological decline in fish stocks over the long term. As captured in Pauly's (2008) paper concerning the underestimation of overfishing, persistent overcapacity leads to systemic ecological change that becomes engrained in our understanding of how these fisheries are affected. His concept of 'shifting baselines' argues that scientific assessments of fisheries reflect the present rather than the past. As a consequence, we maintain a 'gradual accommodation of the creeping disappearance and inappropriate reference points for evaluating economic losses resulting from overfishing, or for identifying targets for rehabilitation measures' (Pauly, 1995: 430). Similar notions of underestimating ecological change as a result of human intervention through fisheries from a longer historical perspective were addressed by Roberts (2010) in his book *The Unnatural History of the Sea*.

The implications of long-term systemic change in the world's oceans have been researched extensively over the last two decades. Perhaps most dramatically this includes Myers and Worm's (2003) highly controversial, and largely debunked, assertion that we will experience a loss of 90 per cent of top predators by 2050. This idea builds directly on Pauly et al.'s (1998) paper, entitled 'Fishing down marine food webs', whereby fisheries are expected to be gradually removing higher trophic level species from the seas and in doing so shift to lower trophic species. He predicts that if fisheries continue this trend, the world will be left with 'jellyfish burgers' as we gradually change the ecological structure of marine food webs over time. These predictions have received considerable attention largely because of their highly powerful message. This has been taken up by several NGOs and science writers to draw attention to the plight of the world's fisheries; some liken the decline to the removal of top predators, such as lions, from the savannah of East Africa (e.g. Frank et al., 2005).

One of the major contributions of the Sea Around Us project has been to focus attention on the uneven geography of effort allocation and overfishing. Fisheries are ever more global in scale, as demonstrated by the already mentioned role of distant-water fishing fleets and the identification of the high seas and deep-sea trawl fisheries as the next frontiers of fishing pressure and decline (Puig et al., 2012). However, the increase of fishing pressure remains a predominantly regional phenomenon based on the expansion policies of both European and Asian fleets. So although fishing is global, there do appear to be identifiable patterns regarding the geography of where fishing occurs.

Reconstructions of fishing effort since 1950 have revealed the spatial variation of overfishing. A series of spatial modelling exercises have indicated that as fishing effort has expanded in and out of Europe and Asia, the overall biomass of fish stocks has declined. As demonstrated by Swartz et al. (2010), the expansion of fishing effort globally has led to a disjuncture between effort and productivity. By calculating the primary production required to support a fishery, and comparing this to the actual primary production of a given area, they assessed how the production frontier has been overrun across the world's oceans since the 1950s. Their assessment demonstrates a longitudinally southward expansion of fisheries from Europe and North America to East and

Southeast Asia, with the most significant expansion occurring in the 1980s and early 1990s. By providing this longitudinal perspective based on a reconstruction of fishing pressure and primary productivity, they stress where and how global ecological limits are overshot. They also show that this expansion is far above agriculture production; while the area under cultivation increased 10 per cent over the same period, fisheries required a 400 per cent increase in exploited areas to reach the same kind of production increases.

What therefore emerge are global hotspots of overfishing, where layers of policy and mismanagement can be overlaid. For example, the high degree of overfishing in Europe reflects the troubled history of the EU Common Fisheries' Policy. The story is complicated, to say the least, but can be roughly divided into debates around allocating catch shares, the political economy of maintaining fishing rights, perverse subsidies keeping fleets inefficient, the failure of bans on discarding fish at sea and the promising emergence of more decentralised modes of devising harvest control rules through regional advisory committees (e.g. Symes, 2012). However, the transnational scope of European fisheries' policy, represented by economic and fishery partnership agreements, has received relatively less attention.

The 'other' hotspot is Asia and the Western Pacific – which represents 72 per cent of total global production, making it by far the largest producing region of the world (FAO, 2014). The history of overfishing in Asian waters is in many ways even more complex than in Europe, given that fishing effort was expanded as much through foreign aid as through expanding domestic interests. The impact has been substantial, as it is estimated that fisheries resources in Southeast Asia have been fished down to 5–30 per cent of their unexploited levels – in the Gulf of Thailand alone biomass is estimated to be only 8 per cent of 1965 levels (Salayo et al., 2008). While it remains unclear whether overcapacity is an issue in small-scale inshore fisheries (Béné et al., 2010), there is broader agreement that the larger-scale offshore vessels need to be better regulated. Although in Malaysia boat licensing has been useful in managing capacity, in most of the other Southeast Asian countries capacity appears to be increasing (Stobutzki et al., 2006). For example, in 2008 the minister of fisheries in Indonesia called for a 300 per cent increase in fishing capacity in 2010 alone, with no details beyond increasing the industrial tuna fleet by 500 boats (Salayo et al., 2008). As in other parts of the world, it is state subsidies that have driven the expansion of these fleets – in the first instance through development aid, followed by local politicians seeking to support coastal communities. In the case of international cooperation, with Japan in the lead, support often came with self-interest in maintaining fish exports to seafood-hungry markets.

Another primary driver of increased fishing effort has been the incursion of fishing vessels from other nations – either legally or illegally. Illegal access is a general problem for many developing countries, due to the presence of unregistered vessels, poor border control and conflicts with inshore fisheries (e.g. Pomeroy et al., 2007). Legal access for vessels from distant-water fishing nations in the global North (Japan, the EU, North America and the former Soviet Union) has been a way for developing countries to extract rent from fishery resources in their EEZs, which otherwise would be difficult due to the lack of capital and investment in domestic fleets. However, effective management of licensed foreign vessels has often proven problematic, because access agreements have been directly tied to development aid and other side payments (Yeeting et al., 2017). Also, licensing fees in the case of small-island developing states form up to 40 per cent of their gross domestic product (GDP) (Havice, 2010). Distant-water fishing nations wanted to maintain historical 'fishing rights' and ensure an adequate volume of fish for their domestic markets, but they have also sought to maintain a level of political influence in specific regions. Coupled with very weak information systems over the status of stocks and domestic aspirations for fleet development, the expansion of access agreements for distant-water fishing vessels only further expanded fishing effort.

But while distant-water fishing nations from the 'traditional North' have been highly influential in expanding effort in the waters of developing countries and driving global catches, their overall effort has been on the decline since the mid-1990s. As shown by Gagern and van den Bergh (2013), the decline in the catches of these nations has been taken over by low-middle-income countries and emerging economies in Asia and Africa (Figure 3.1). While China, Taiwan and South Korea appear to have levelled off their catches, Southeast Asian countries have continued to increase

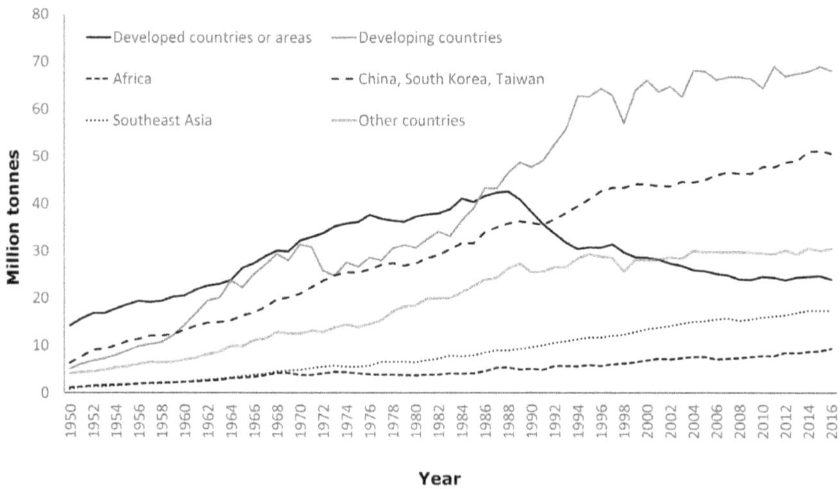

Figure 3.1 Landings of marine fisheries for five clusters of countries.

Source: adapted and updated from Gagern and van den Bergh (2013).

Note

'Other countries' excludes Peru.

the amount of fish landed. Driven by a process of gradually less efficient fisheries, this shift from the North to the South indicates a potential shift in who is driving the global expansion of fisheries by investment in fishing effort. While further domestication of fishing effort has boosted food security and the GDP of developing countries, it also has implications for the sustainability of fisheries. As noted by Gagern and van den Bergh (ibid.), the development partnerships driven by the EU, but also to different extent by the US and Japan, have set onerous conditions on developing countries to earmark access fees to specific investments around fisheries management. The emergence of alternative distant-water fishing nations, asking only for 'pay–fish–go' access agreements with no further conditions, has higher economic and political appeal. But given the poor performance of these emerging distant-water fishing nations on IUU fishing in their domestic waters, governing excess fishing pressure through South–South access agreements may prove even harder.

The expansion of fisheries has seen diminishing returns and pushed the production frontier back to open up new production areas. The global distribution of overcapacity and overfishing has gradually changed over time, moving from the industrialised global North to the developing global South – perhaps because of diminishing returns from a large number of fish stocks. This appears not to have reduced fishing effort, but instead shifted effort to emerging economies looking to expand beyond their own, often overexploited, coastal waters. But as we now explore, catching fish from the wild is not the only means of maintaining increasing production. And the apparent decline in fishing capacity from the global North may only foreshadow a shift from fishing vessels on the water to the more efficient 'capture' of seafood through global trade.

Geographies of aquaculture production

With the diminishing returns of global capture fisheries, the only reason that global seafood supplies have continued to grow since the 1980s is use of aquaculture. Globally the sector has been growing at an average annual rate of 6–7 per cent (FAO, 2014), and now contributes half of the fish produced for human consumption. In some countries, mostly in Asia, growth rates have been even higher. For example, because of the expansion of pangasius catfish, Vietnam has posted an annual growth rate of around 30 per cent a year – making it the fastest-growing food sector in any country (ibid.). In India and China, the world's two most populous nations, aquaculture now supplies two-thirds of all aquatic organisms produced. It also provides 13 per cent of the world's animal protein, excluding eggs and dairy (calculated from FAO, 2012; World Bank, 2012).

Aquaculture comes together with capture fisheries around the same goal of supplying fish protein, but they are different in nearly all other ways. Whereas management and conservation of wild stocks underlie debates around the future of capture fisheries, aquaculture aligns with a strong 'productionist' agenda, where increasing the efficient supply of fish remains imperative (e.g. Barton and Fløysand, 2010). Captured by the 'blue revolution' tagline, growth and expansion in the aquaculture sector are thought to be limited only by technology. From this perspective, the trajectory of aquaculture has been comparable to that of animal husbandry, except that the domestication of fish species has occurred over a much faster timescale. In simple terms, the major barriers to advancing aquaculture production first lay in transitioning systems that rely on wild-caught fish 'grown out' in ponds or cages to closed reproductive cycles through artificial breeding. As systems look to increase the intensity of production, the next barrier is to improve efficient feeding and, directly related, to ensure waste removal to maintain water quality. Proponents of the blue revolution have focused on increasing the production in aquaculture to contribute to global food security, thereby reducing the pressure on wild stocks through diminished reliance on fishmeal-based feed for carnivorous species while increasing the availability of high-value species such as salmon and shrimp.

Despite its phenomenal growth rate, the global aquaculture industry produces a surprisingly narrow set of species. The FAO (2016a) lists 580 species and species groups in its global production database, but a relatively small number forms the bulk and value of production (see Figure 3.2). Nearly

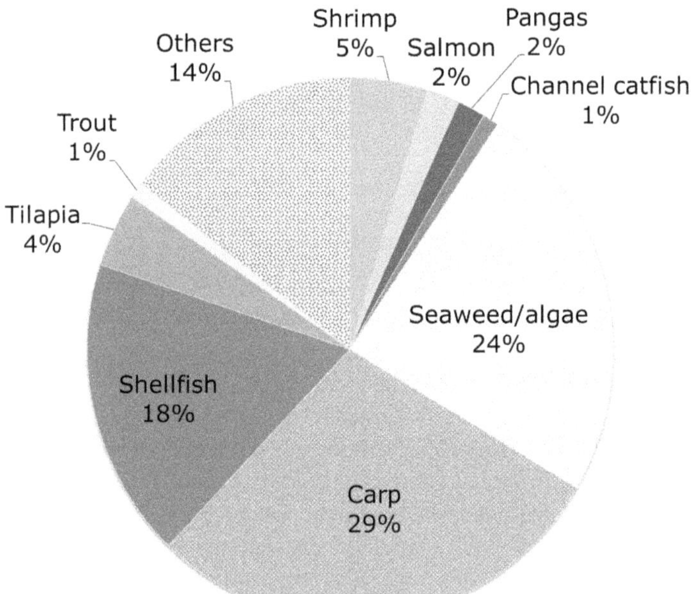

Figure 3.2 Contribution of key crops to total global aquaculture production by volume, 2014.

Source: modified from FAO (2016).

three-quarters of this production is made up of seaweed, shellfish and the broader category of freshwater carps. The next two most abundant species by volume are tilapia and shrimp, followed by salmon and pangasius, and then channel catfish and trout. Those species lumped together under 'other' are a combination of mostly highly specialised and high-value species (including molluscs, tuna, eel and flatfishes) and a range of other aquatic organisms produced mainly in rural economies of the global South (such as frogs).

Taking a closer look, aquaculture is mostly a tropical and inland exercise. It is also very Asia-centric. Global bivalve production is made up predominantly of clams, oysters, scallops and mussels, with 87 per cent grown in aquaculture systems (FAO, 2016b). China is the top bivalve producer in the world, accounting for 70 per cent of global production. The remaining 30 per cent is spread between other Asian and European countries, including Japan, Thailand, the Philippines, Spain, the Netherlands and France. In the case of seaweed there are domestic food markets and also a range of derived products used in food supplements and pharmaceuticals. However, again production is highly skewed to Asia, with China and Indonesia together making up 92 per cent of the global output. Carps fall into the same category, with China, India, Bangladesh, Myanmar, Vietnam and Indonesia making up 96 per cent of production. Despite the spurious accuracy of Chinas statistics, there is little dispute that this country leads global production, but South and Southeast Asia also play an essential role in the production of these species. Shellfish and carp do play an important role in European aquaculture, but Europe remains overall a small player globally.

The growth of certain species and species groups has been dependent on the ease of domestication. Seaweed and mussels are propagated by removing cultivars and spat from the wild and putting them in a more-or-less controlled grow-out system – albeit often controlled areas or attached to hanging ropes. The domestication of the various carp species has a long history in the rice/fish systems of Asia, where fish self-recruit, are trapped in flood waters or are caught and raised in ponds. However, intensification of carp cultivation was only possible when the reproductive cycle was closed and intensive systems developed across the region (e.g. Hishamunda et al., 2009). Shrimp, tilapia and pangasius are similar: domestication was only possible when the reproductive cycle was closed, making production less dependent on wild sources and therefore more efficient and profitable. It is these latter species that have been industrialised for global mass consumer markets. It is also notable that these same species are among the fastest growing globally (Figure 3.3).

Given the strong productionist agenda for aquaculture and the goals set out by the industry for food security in the face of stagnating fisheries' production, the only way forward seems to be to maintain growth in aquaculture. Although historically seen a laggard in global aquaculture production, Africa has a ten-year average growth rate of 11.9 per cent, outperforming Asia and the Americas by almost a factor of two – albeit with half the output

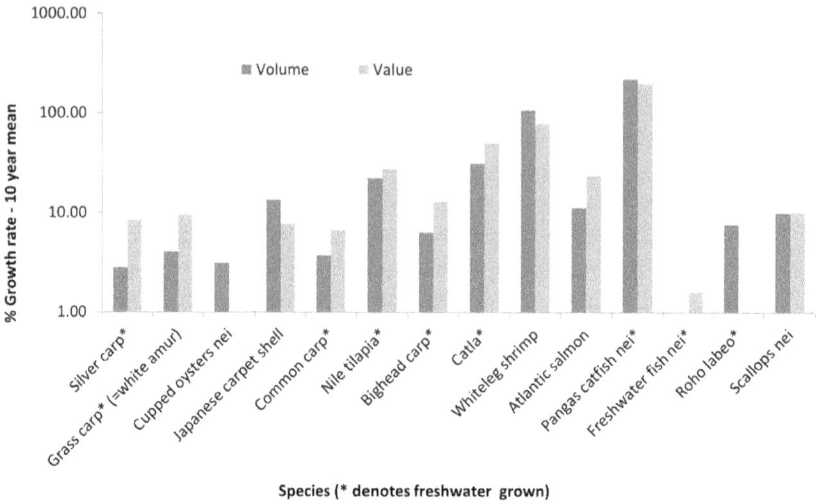

Figure 3.3 Ten-year average growth rates of selected species by volume produced.

Source: data from FAO (2014); adapted from Bostock (2010).

Note

NEI indicates fish species 'not elsewhere included'.

of Europe and 2 per cent that of Asia. Two of the poorest-performing regions of the world are Europe, which has experienced an average ten-year growth rate of only 2.9 per cent, and the United States, which experienced negative growth at –0.7 per cent. Looking back further in history, the growth rate of the EU and the Americas peaked in the 1980s, but by the 2000s had declined by 4.9 per cent and 7.2 per cent respectively (Figure 3.4). Despite continuing to outperform capture fisheries by an average of 6.3 per cent over all continents, the decline in growth rates for aquaculture is concerning when considering the critical role of fish farming in maintaining growth in overall seafood output.

Ownership in the aquaculture industry is vertically integrated to varying degrees. On the one hand ownership in shrimp production is very much spread out among large numbers of small and middle-sized farms and only some large-scale facilities (Islam, 2014). On the other hand ownership within the salmon aquaculture industry has become highly concentrated, with roughly 22 companies controlling two-thirds of the world's farmed salmon production in 2016 (Bush, 2017). Although the salmon fishing industry is made up of many small businesses that operate at arm's length from processing corporations, the salmon farming industry is made up of companies with corporate affiliations. The four largest multinational companies involved in global salmon aquaculture production are Marine Harvest (representing the

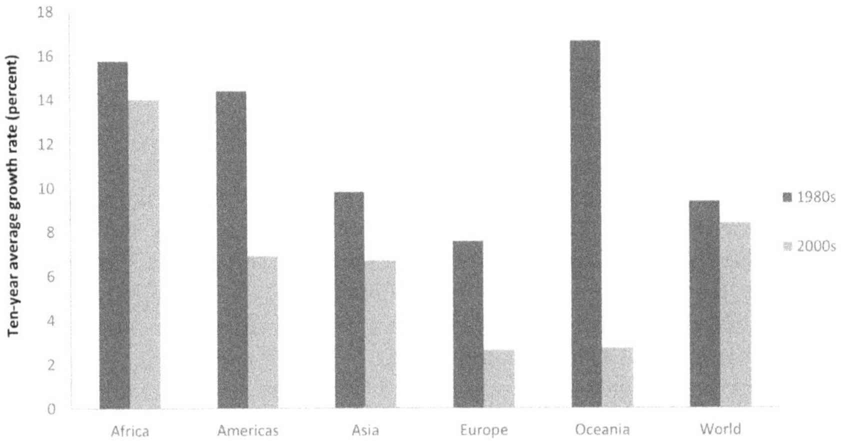

Figure 3.4 Comparison between ten-year average growth rates, 1980s and 2000s, by continent.

Source: based on data from FAO (2012).

merged companies Stolt-Nielson and Nutreco) and Mitsubishi (representing Cermaq and Salmones Humboldt), followed by Leroy Seafood Group and Salmar. As Naylor and Burke (2005: 190) argue, 'it is typical for an aquaculture multinational to have subsidiaries that include feed, hatchery, grow-out, distribution, and value-added processing companies, and most of the multinationals have operations on at least three continents'.

The declining growth is slowly changing the global productionist discourse around aquaculture, based on technology and expansion, into one of 'frontiers'. In its 2011 report entitled *Blue Frontiers*, the WorldFish Center outlined the challenges of reducing environmental impacts to maintain aquaculture production (Hall et al., 2011). The notion of frontiers is important here, given these are seen as barriers to the continued growth of the industry. The use of frontiers stresses that continued expansion is dependent on the quality of the environment in which fish culture is carried out. For some species this focuses attention on negative impacts of environmental withdrawals (protein, water) and additions (nutrients, waste), while for species groups like mussels their interaction is positive – withdrawing and filtering more than they add. Whatever the outcome, the logic of these interactions between environment and aquaculture supports studying and understanding these impacts largely as a system's problem, and linking this with the local and global geography of environmental impact.

Understanding aquaculture as an interconnected system, much like any primary production system, expands the scope of environmental impacts. Using such a systems approach in the late 1990s, a group of Swedish scholars introduced the notion of an ecological footprint for aquaculture production (Kautsky et al., 1997; Folke et al., 1998). Not only did the footprint demonstrate the broader impacts of aquaculture, based on the flows in and out of the system, but it also helped to highlight both the local and the global linkages of fish farming. At the same time, a movement was emerging against aquaculture because of growing awareness of the impact salmon aquaculture was having in North America and Scotland and the impact shrimp aquaculture was having in coastal habitats across the tropics. Most of the initial impacts observed were local. For salmon aquaculture, based on cages along coastal areas, these impacts included eutrophication of bays and seabed fouling, outbreaks of sea lice affecting wild populations and genetic dilution resulting from interbreeding of domesticated salmon with wild salmon (Folke et al., 1994; Asche et al., 1999; Naylor et al., 2003; Naylor and Burke, 2005). The impacts of shrimp aquaculture were by and large more severe, or at least more visible. The poorly regulated expansion of ponds in Latin America and across Asia led to widespread deforestation of mangrove forests, coastal water pollution and soil degradation (e.g. Dierberg and Kiattisimkul, 1996; Kautsky et al., 2000). But while these impacts were severe in their own right, the impact that received most attention was the global impact aquaculture was having on wild stocks of fish used for feed.

The environmental impact of the feed used as input for aquaculture highlighted a central paradox of the productionist agenda that underscored the expansion experienced by aquaculture over the 1980s and 1990s. While aquaculture appeared to be a solution to the decline of capture fisheries, the high levels of fish protein in aquafeeds meant that it was also a contributing factor to the collapse of fisheries' stocks worldwide (Naylor et al., 2000). While the expansion of capture fisheries stagnated, the quantity of fish catches not used for direct human consumption remained relatively constant at around 21 million MT per annum (FAO, 2016b). Of this one-third goes to aquaculture feed (Metian, 2009). The share of fishmeal used by aquaculture has continued to increase, from 8 per cent in 1988 and 35 per cent in 2000 to 45 per cent in 2005 and over 50 per cent in 2010 (Hall et al., 2011). There are substantial differences between the amounts of fish protein used in aquafeeds for different species, measured as a fish-in/fish-out ratio. In the mid-1990s this ratio was overall 1.04 kg of wild-caught fish for every 1.00 kg of aquaculture fish produced – with some species like salmon and shrimp around three to one (Naylor and Burke, 2005). By the late 2000s the overall ratio was down to 0.63, reflecting in part the expanding volume of omnivorous fish produced on farms as well as market pressures to reduce fishmeal and fish oil levels in aquafeeds (Naylor et al., 2009). However, carnivorous species groups such as salmon and shrimp remain higher, at 1.3 and 1.7 respectively. And even tilapia and carp stay above 1.5 because of the higher efficiency of using fishmeal and fish oil in feeds.

The use of fishmeal and fish oil also expands the geographical footprint of aquaculture. Overall Peruvian anchovy fisheries provide half of global fishmeal and around a third of fish oil (Fréon et al., 2014). More than half of Peru's and Chile's fishmeal exports now go to three Asian countries: China, Japan and Taiwan. Asia is the largest global consumer of fishmeal overall (57 per cent), with China accounting for half of the regional total (as calculated by Naylor et al., 2009). However, aquaculture-producing countries such as Thailand, Chile and China are consuming increasing amounts of domestic fishmeal supplies while exporting lower volumes of feed, and importing fishmeal from other countries (Deutsch et al., 2007). As China increases its already considerable production, concerns are growing over the supply of fishmeal from coastal fisheries with weak management (Cao et al., 2015). Using Norwegian salmon and Thai shrimp as examples, there appears to be no wholesale movement of fish from developed to developing countries. For Norway supply is either regional or from Peruvian fishmeal, while in Thailand Peruvian supply is supplemented with fishmeal from both Northern Europe and East Asia. But as Naylor et al. (2009: 15104) argue, perhaps the most concerning overall aspect of these new global patterns of sourcing fishmeal is the 'rift' that is emerging 'between the economic performance of aquaculture producers and the impacts that the industry has on the marine ecosystems that supply them'.

The loudest critics of the use of these fish catches for fishmeal regard them as the ultimate form of 'fishing down the food web' given that the catch supplies export-oriented aquaculture systems and does not (necessarily) contribute to local or global food security. To address this problem, Naylor et al. (2009) recommend instead to 'farm down the food chain'. By moving to omnivorous species and introducing new plant-based feeds (e.g. soy), there is an opportunity to relieve pressure on wild stocks. As indicated by the overall decline in fishmeal use, in this respect progress is being made. Overall, three-quarters of aquaculture production is of low trophic species, largely because of the dominance of freshwater fish, representing 85 per cent of total production (Tacon et al., 2009). But there is also an emerging countertrend tending towards aquaculture farming up rather than down the food chain. As Tacon et al. (ibid.: 99) argue,

> despite having a more ecologically balanced trophic pyramid structure … aquaculture is currently set out towards the expansion of high trophic level and higher market value finfish species, particularly within developed countries, and to a lesser extent developing countries and China.

Within developed countries, more than 90 per cent of total finfish production involves high trophic level species. In contrast, between a quarter and third of total finfish production in developing countries, including China, is high trophic level species. This trend is increasing.

In recent years attention has been given to the systemic environmental impacts involving species that are traditionally thought to be less environmentally damaging. For example, carp are thought to have a lower environmental footprint because they are omnivorous. However, when looking beyond feed the sheer volume of carp produced means that issues such as eutrophication and acidification, and even energy use and biodiversity, may still become critical concerns (Hall et al., 2011). These impacts are not only distributed between species but also geographically. For example, Hall et al. (ibid.) show that aquaculture production is markedly more efficient in salmon-producing nations regarding overall environmental impact per MT produced – except biotic depletion due to fishmeal inputs. Other patterns are difficult to distinguish, other than China being less efficient concerning energy demand, acidification and climate change potential while Indonesia and the Philippines score higher regarding eutrophication. Inland fish perform similarly, but US and Vietnamese catfish are least efficient. Overall, however, reaching a global production level of 100 million MT by 2030 is estimated to increase environmental pressure by a factor of between 2.0 and 2.5 for all the categories they studied.

When taking into consideration these broader environmental impacts, yet another geography appears to be emerging. The results of the Hall et al. (ibid.) study show that in 22 of the 36 environmental indicators the best performers have a 50 per cent lower environmental impact than the worst performers. They make two basic observations. First, despite the dependence on lower trophic level species, Asia is emerging as a region of environmental concern because production is set to double by 2030. This result is supported by the Global Aquaculture Performance Index, which indicates that Asian countries account for the 15 lowest species/country scores (Volpe et al., 2013). Second, the 'salmon countries' appear to be demonstrating far better environmental performance, with the exception of feed. The reasons for this vary, but generally relate to economic competition that is driving production, industrial consolidation and vertical integration, as well as a generally higher level of environmental regulation in the countries where salmon is grown (Tveteras, 2002). The question is whether a 'salmon model' is a viable option for regions such as Asia. In some respects, similar patterns are already visible in other export species such as shrimp, (Vietnamese) pangasius and tilapia. But such models are less apparent in the bulk production species such as carp, which (except in China) remain primarily rural production systems.

The above paragraphs emphasise that different aquacultural geographies emerge when taking into account production, feed and environmental impact. Despite aquaculture being widely touted as a global industry and contributing to the provision of global animal protein, these geographies are instead partial. The flows of inputs and outputs to and from different aquaculture production systems have different ecological footprints. Some are local, some are regional and some are transnational without being global. The productionist agenda of aquaculture sets out a very different solution orientation for these limits to the blue revolution than for capture fisheries. Hall et al. (2011) are again instructive here. In overcoming what they term 'blue frontiers' to aquaculture, they focus on the role that technological innovation, regulation, monitoring and compliance, and supply and demand analysis can play in different continents. Noteworthy is their distribution of these four areas across regions. All require attention into the future, but the two areas that were consistent in needing investment and development are technology innovation and regulation. The category receiving the top ranking for immediate attention is strengthening regulatory capacity in East and Southeast Asia and South America.

The attention given to technological innovation is of particular interest because it represents a shift away from pushing higher levels of production and towards ensuring that increases in production make more efficient use of land, water and feed resources. Framed as a renewal of the greening of the blue revolution or ecological aquaculture (Costa-Pierce, 2010), a series of new production systems have emerged that attempt to reuse or recycle nutrients within the systems. The diversity of these systems is growing, but they coalesce around three types – integrated multitrophic aquaculture, recirculation aquaculture systems and aquaponic systems which integrate fish with hydroponic systems (see Table 3.1 for details). Despite the hope that these systems bring for moving the sector to more efficient resource use, they also

Table 3.1 Comparison of new-generation aquaculture systems

System	*Key characteristics*	*Key critiques*	*References*
Integrated multitrophic aquaculture	Integrated aquaculture systems where by-products and waste flows are utilised by other cultured species	High capital cost Unclear economic benefits from different species mixes	Barrington *et al.*, 2009; Troell *et al.*, 2009
Recirculation aquaculture systems	Near closed-loop circulation of waste flows through filters and then recirculating back into the tanks to save water and reuse nutrients	High capital and unit cost High energy usage	Martins *et al.*, 2010; Badiola *et al.*, 2012
Aquaponic systems	Integrated aquaculture systems with hydroponic vegetable production and recirculation of water and nutrients	High unit costs High energy usage	Blidariu and Grozea, 2011; Love *et al.*, 2015

face some critiques. First, the commercial use of these systems has been slow, prompting questions about whether they are a feasible option over the long term (Bunting, 2013). The complex nature of recirculation aquaculture systems, for example, and their high operating costs have led to more failures than successes despite having been developed over 25 years ago (Bostock et al., 2010). There are also trade-offs between environmental outcomes. For example, aquaponics might create efficiencies in water use in arid climates, but have a higher cost and higher energy unit of production than pond systems (Edwards, 2015).

So while these systems represent some of the frontiers of aquaculture technology, they also face limitations to becoming commercially viable alternatives. One of the key reasons why they have not become economically viable is because of competition and weak internalisation of environmental and social externalities into the cost of production. As long as the economic cost of aquaculture production remains at the level of the farm, and does not consider the environmental and therefore societal benefits of these production systems, they will remain uncompetitive (ibid.). And as long as the cost of these systems remains high, questions will continue to be asked over the accessibility of the systems beyond wealthy OECD countries. If these systems can become viable, as is gradually being seen in highly industrialised sectors such as Vietnamese pangasius production, environmental benefits may be realised (e.g. Ngoc et al., 2016). But if the cost of these systems stays too high, they will remain simply a means of increasing the competitiveness of aquaculture in regions such as Europe and North America that are currently less competitive in the global market. As Edwards (2015) outlines, the future of aquaculture may instead rely on sustainable intensification using low-level technologies such as feed, fertilisation and aeration of existing systems in regions where production has been and continues to be highest.

Social geographies of seafood sustainability

The previous sections provide an overview of the major issues raised around seafood sustainability. The sustainability issues raised are seen most commonly as technical, relating to both the technologies and the biophysical ecology of fisheries and aquaculture. But these issues are also fundamentally social because they are continually redesigned by human actors; the decisions that fishers and fish farmers make day to day and more strategic decisions about dimensions of production are embedded in broader networks of government policymakers, industry partnerships, NGO pressure groups and both local and global markets.

The social geographies of seafood production embed the decisions of producers into globalised social relations that have considerable influence over whether and how the above issues are overcome. Recall that our interpretation of globalisation highlights the intersection of the space of flows and the space of place, which draws out both social networks and their intersection with biophysical processes of seafood production. Recall also that central to the dissection of the space of flows and the space of place is an understanding of who controls the flows of information, regulation and materials – the so-called switchers and programmers of global networks (Castells, 1996). Each of the sustainability issues covered here can be analysed in these terms, highlighting who does or does not contribute to perpetuating a sustainability issue, and who is (un)able to help its resolution. Such questions then build different 'social geographies' because they highlight relations of production in different socio-spatial configurations (flows, networks and places) and their influence over where sustainability issues are dealt with and by whom (Massey and Allen, 1984). Identifying what these social geographies are and how they relate to fisheries and aquaculture production is a crucial step in understanding the relevance and effectiveness of the different governance arrangements outlined in detail in the next chapters.

The following subsections provide an overview of the major 'social geographies' that emerge from the literature on sustainable seafood: global and local interconnections, public and private forms of regulation, inclusion and exclusion of social actors, and divergence and convergence of problems and solutions.

Global and local

The global–local divide is a critical social geography, in that it helps to determine analytically the extent to which local processes, embedded in the space of place, are influenced by global flows. The review of fisheries and aquaculture primarily highlights the distribution of sustainability concerns between regions and countries around the world. But each of these issues also exists at the intersection of flows and places, and the outcome can lead to 'frictions' (Tsing, 2011). Such frictions lead to new creative results that combine the local and global and thereby empower producers to do things more efficiently or more sustainability or alternatively lead to a universal rolling out of an idea or mode of production that disempowers local producers to respond to the sustainability challenges facing them.

Many of the sustainability challenges listed above for both fisheries and aquaculture can be seen through this global–local lens. The perceived global nature of fisheries is observed most clearly through transboundary fish stocks, which by definition straddle political boundaries. Many other species are less global, meaning they cross no more than one international boundary. But regardless of this nuance, the point remains that the movement of these fish has led to tensions between the interests, economies, cultures and livelihoods of fishers in different parts of the world. It is not possible to illustrate the full diversity here; instead we highlight that the more political boundaries fish cross, the more complex the social realities of the fishers targeting them and the forms of regulation that define them as property (Havice and Campling, 2010). Behind many of the decisions on how access to fish is distributed are issues of national economic value. In addition, it is not only the fish that 'flow' around the world but also the capital used to finance the fishing industry. The effect of such finance is key to understanding sustainability, as fishers are constrained in their ability to fish more sustainably by the expected returns on capital for investors (Constance and Bonanno, 1996; Havice and Campling, 2017).

Aquaculture is different to fisheries in that the fish grown are sedentary – constrained by the nets, ponds or pens in which they are produced. But a fundamental difference is that in most cases where an aquaculture system is introduced it involves new technologies developed in different parts of the world, which in turn leads to the new allocation or use of local natural resources (such as land and water). The industry is also increasingly based on international and often global networks of actors, so aquaculture is most commonly introduced through the global flow of technology, knowledge and capital. Nearly all of the industrialised aquaculture sectors have been based on the introduction of knowledge and technologies from other countries, either through development aid from the 1970s to 2000s or foreign direct investment since (Bailey et al., 1996; Belton and Little, 2011). The expansion of shrimp aquaculture, for instance, was based on a series of waves of spawning, feed and out-grow technology extending from Taiwan, the US, the EU, Thailand and more recently China to various parts of the often 'developing' world (Lebel et al., 2002). Similar waves of innovation are seen for other major species, such as tilapia, salmon and pangasius (Barton and Fløysand, 2010; Bush and Belton, 2012). The arrival of these species and production from the 'global to the local' has therefore been associated with social and environmental changes that are reasons for sustainability concern – whether it be the introduction of exotic species, diseases or the social consequences of boom economies associated with their introduction (Hall, 2004; Edwards, 2015). Coupled to this is the fact that because aquaculture is 'new', wherever it starts local (state) regulation is often not in place.

Seeing seafood production with respect to the tensions and opportunities that emerge through the intersection of the global and the local is key to understanding the challenges of governing many if not all of the sustainability issues outlined in this chapter. The frictions that emerge out of this global–local interaction do not necessarily hold negative consequences; they can open up opportunities for sharing and innovation that can lead to sustainability solutions. In the following chapters we return to how the various attempts by the sustainable seafood movement to push towards sustainability deal with the complexity of global–local interactions, either by reifying the position of a perceived global imperative over local interests or carefully and successfully negotiating between the global and the local.

Public and private

The distinction between public and private spheres of ownership and regulation also runs across the sustainability issues outlined earlier. In both fisheries and aquaculture, a better understanding of the boundaries between public and private resources, credibility and authority is central to comprehending how these sustainability issues can be dealt with over the long term.

Common to both fisheries and aquaculture are questions over the boundaries of public and private resources. For fisheries, there are questions over the ownership of the fish stocks. While it is undisputed that these fish stocks are common property, there are sharp differences in opinion over ownership of the rights to exploit these stocks. As we go to detail in the next chapter, western liberal economies have pushed for ITQs, with the assumption that private allocation of these catch shares avoids the so-called tragedy of the commons – fishers putting their own return from exploitation above the shared stewardship of the resource (Hardin, 1968). In aquaculture the debate over the public–private divide is less pronounced, given the assumption that the boundary of a farm delineates private ownership of the production system. But in practice many of the sustainability issues related to aquaculture extend beyond the limits of farms to 'public' water, land, forest and biodiversity resources (Kautsky et al., 2000; Bush et al., 2010). The delineation of private ownership subsequently frames impacts of production on water resources, habitats and biodiversity as externalities to be dealt with by public regulation.

It is clear that through resource ownership and environmental impact the notion of public and private regulation is also blurred. The assumption that state regulation has been unable to deal adequately with the direct management of fishery resources, either within a sovereign jurisdiction or through RFMOs at the international scale, is argued throughout the fisheries management literature (e.g Bavinck et al., 2005; Allen et al., 2010; Cullis-Suzuki and Pauly, 2010; Khalilian et al., 2010). Likewise, the poor or non-existent regulation of environmental externalities is seen as a failure of the state. Central to this book is the observation that this perceived regulatory gap has been 'filled' by a wide range of private tools and arrangements aimed at directly setting rules for the management of these externalities, or pushing (in a variety of ways) for improved state regulation.

The public/private geographies of sustainability, in terms of both ownership and regulation, are a central challenge for understanding the success of the sustainable seafood movement. As argued in the previous chapter, the networked constitution of the movement means that both public and private actors are collaborating to find solutions to the problems faced by the fisheries' and aquaculture sectors. These actors are experimenting with the limits of their legitimacy in setting rules, the credibility of the sustainability claims they can make over the performance of the sector, and ultimately their authority to gain regulatory control of both private and public resources, actors and institutions.

Inclusion/exclusion

Our review demonstrates the distribution of sustainability issues evident in aquaculture and fisheries between developed and developing countries.

Inherent to this geography is a range of concerns about social equity and environmental justice. The literature on fisheries and aquaculture has a long tradition of examining how different groups of producers are either included or excluded regarding access to the industry, being marginalised through environmental degradation or being excluded from livelihoods based on these industries. In doing so, implicit or explicit divisions are commonly made between large and small-scale producers.

Social scientific research and policy on fisheries have tended to focus on the impact of expanding fishing fleets, IUU fishing and the decline of fish stocks on small- and large-scale fishers. For instance, the expansion of the fishing industry in many parts of the world has often come at the expense of small-scale fishers, because of either their poor efficiency relative to larger vessels, or the sale of quotas or vessels as part of either voluntary or forced consolidation of the industry, or their complete exclusion from quota allocations (see for example Maurstad, 2000; Isaacs, 2006; Salayo et al., 2008; Yandle and Dewees, 2008; Hara and Njaya, 2016). Their exclusion is linked to the allocation of fishing rights to formally registered fleets, and sometimes foreign(-owned) vessels, often leading to the overexploitation of the coastal resources on which they depend (e.g. Alder and Sumaila, 2004; Atta-Mills et al., 2004; Berkes et al., 2006; Seto et al., 2017). IUU fishing policy is also seen to affect small-scale fishers unfairly because the formal systems for accounting for their actions, including state regulation, are weak or non-existent (Hauck, 2008; Castello et al., 2009). Social scientific research has tended to focus on the small-scale sector because these issues are pertinent to the livelihoods of the more than 12 million (estimated) small-scale fishers worldwide (Pauly, 2006). At the other end of the spectrum, economists have in general terms tended to focus on large-scale fisheries, exploring issues of effort, quota allocation and policy effectiveness. But this focus has often come at the cost of looking into what might be classified as the missing middle of fisheries. The reasons for this apparent oversight are understandable, given that there is relatively scant information available on these fleets.

A similar picture is found in the aquaculture industry. As outlined by Belton et al. (2017a), there are widespread assumptions about the predominance and desirability of 'small-scale', low-intensity farming technologies that contribute directly to livelihood security in developing countries (e.g. Bondad-Reantaso and Subasinghe, 2013), while environmental degradation and social dislocation are associated with the rise of 'industrial' (often export-oriented) aquaculture in compromising the food security of communities in Southern producing nations (e.g. van Mulekom et al., 2006; Nayak and Berkes, 2011). But despite the dominance of these narratives and their influence in shaping science, policy and popular perceptions, Belton et al. (2016, 2017b, 2017c) argue that they do not adequately represent the current diversity of aquaculture in developing countries, its aggregate 'macro' effects on food security or, by extension, the distribution of sustainability concerns. Instead, they argue, aquaculture should focus on the 'quiet revolution' in farmed fish production which involves intensive systems owned by a range of producers in terms of 'scale' or 'size' and selling into growing domestic markets.

The consequence of this mischaracterisation of fishers and fish farmers is that governance solutions can be misdirected and, as a result, less effective. By making assumptions about who is responsible for poor environmental or social performance, a priori choices are made about who is included in or excluded from contributing to their resolution. In the following chapters we assess how the various tools and approaches that come under the sustainable seafood movement are inclusive of a broad range of producers. In particular, what are the consequences of inclusion and exclusion for the effectiveness of the movement in pushing a global sustainable agenda, and what effects do decisions over who is included or excluded have on the credibility of those organisations pushing this sustainability agenda?

Divergence and convergence

The final social geography that emerges is the divergence or convergence of sustainability issues, technologies and practices in global seafood production. The review of fisheries and aquaculture offers a picture of considerable diversity across the globe, but it also provides some indication that the sustainability issues in fisheries repeat themselves across time and space. Butcher's (2004) notion of pushing back the frontier is a clear expression of this extension of sustainability issues in fisheries, indicating that technology has enabled fishers to exploit resources further out to sea and deeper below the sea surface over time. A similar metaphor is possible for aquaculture as well. As aquaculture technologies have diffused around the world, they have brought with them the same sustainability concerns.

But, related to the other geographies, it is unclear whether such a picture of convergence is as clear-cut as one might think. The local diversity of environments, cultures, politics and economics also brings substantial diversity in how both fisheries and aquaculture play out, despite the apparent diffusion of technologies used to harvest or grow fish. As St. Martin (2005) observes, assumptions of divergence can lead to policy categories that undermine the emergence of divergent understandings of what fisheries are. He argues that developmental discourses of 'first world' and 'third world' fisheries have overemphasised the convergence and indeed homogeneous character of fishing practices and modes of production. Third world fisheries are framed through common resources, community, participation, cooperation and prescribed 'co-management'. Conversely, first world fisheries are characterised in terms of a capitalist mode of production and the consequently prescribed privatisation through ITQ. The consequence, St. Martin argues, is that the diversity of fisheries in regions like the US northeast is better characterised in the prevailing 'third world' rather than 'first world' discourse, forcing a rethinking of categorical assumptions and a return to different sets of problems and solutions.

Because aquaculture is perceived as a technologically driven sector, it is subject to similar notions of convergence – often, as outlined above, under the banner of the 'blue revolution'. However, the expansion of aquaculture has also undergone simultaneous movements of convergence and divergence in regions such as Southeast Asia. As Hall (2004) notes, Thailand, the Philippines and Indonesia all began with shrimp aquaculture in the 1970s, but had subsequently very divergent development paths based on the role of development aid, state support to multinational companies and their response to subsequent rounds of disease and environmental degradation. When analysed through these simultaneous processes, the divergence of Thailand as the primary producer in the world by the 2000s can be explained by more than the advancement of technology alone. Like fisheries, then, categorical assumptions about how fish production is practised and what the potential solutions are often prove problematic.

The question for the sustainable seafood movement is how sensitive the tools and arrangements through which it operates can be given the tendency to make assumptions based on the convergence of production practices. Are countries and sectors moving together, or further apart? This question is addressed in the following chapters along with a related challenge for the sustainable seafood movement itself: is it also coming together to form a coherent global movement, or is there a divergent trend in developing different modes and practices in related sustainability movements that correspond to the diversity of challenges faced by the industry?

Conclusion

While seafood is considered a coherent sector, it is made up of complex global geographies of production, trade and consumption. The first level of variation in the seafood sector is between aquaculture and fisheries. These two sectors not only represent fundamentally different modes of production but also very different geographies of sustainability impacts and solutions. But perhaps most importantly, the second level of variation lies within the social geographies of seafood production. While these social geographies create even greater complexity, they also represent a linchpin for orienting the sustainable seafood movement to steer the industry to the sustainable provision of fish protein supply. In this sense, the seafood movement is as much a movement of social steering as it is of technological intervention.

Highlighting both the technical and the social geographies of seafood production makes it clear that technological advancement is but one dimension of sustainability in the sector moving forward. The following chapters build on this argument by exploring the strategies, tools and governance arrangements that fall under the sustainable seafood movement, and questioning the extent to which they can deal with this apparent diversity and complexity. In short, the fact that this diversity is mostly hidden from consumers in the markets has driven the sustainable seafood movement but has also complicated its effectiveness. We are interested in how the various sustainability challenges and social geographies outlined in this chapter are reflected in the assumptions made in the design of these strategies, tools and governance arrangements. Building on this analysis, the next two chapters explore the diversity of public and private forms of global seafood regulation.

Note

1 See https://ec.europa.eu/fisheries/cfp/international/agreements_en (accessed 18 August 2017).

References

Agnew, D. J., J. Pearce, G. Pramod, T. Peatman, R. Watson, J. R. Beddington and T. J. Pitcher (2009). 'Estimating the worldwide extent of illegal fishing', *PLoS One* 4(2), e4570.

Alder, J. and U. R. Sumaila (2004). 'Western Africa: A fish basket of Europe past and present', *Journal of Environment & Development* 13(2), 156–178.

Allen, R., J. Joseph and D. Squires (2010). *Conservation and Management of Transnational Tuna Fisheries*. London: Wiley-Blackwell.

Anticamara, J., R. Watson, A. Gelchu and D. Pauly (2011). 'Global fishing effort (1950–2010): Trends, gaps, and implications', *Fisheries Research* 107(1), 131–136.

Asche, F., A. G. Guttormsen and R. Tveterås (1999). 'Environmental problems, productivity and innovations in Norwegian salmon aquaculture', *Aquaculture Economics & Management* 3(1), 19–29.

Atta-Mills, J., J. Alder and U. Rashid Sumaila (2004). 'The decline of a regional fishing nation: The case of Ghana and West Africa', *Natural Resources Forum* 28(1), 13–21.

Badiola, M., D. Mendiola and J. Bostock (2012). 'Recirculating aquaculture systems (RAS) analysis: Main issues on management and future challenges', *Aquacultural Engineering* 51, 26–35.

Bailey, C., S. Jentoft and P. Sinclair (1996). *Aquaculture Development: Social Dimensions of an Emerging Industry*. Boulder, CO: Westview Press.

Barrington, K., T. Chopin and S. Robinson (2009). 'Integrated multi-trophic aquaculture (IMTA) in marine temperate waters', in D. Soto (ed.) *Integrated Mariculture: A Global Review*, Fisheries and Aquaculture Technical Paper No. 529. Rome: FAO, 7–46.

Barton, J. R. and A. Fløysand (2010). 'The political ecology of Chilean salmon aquaculture, 1982–2010: A trajectory from economic development to global sustainability', *Global Environmental Change* 20(4), 739–752.

Bavinck, M., R. Chuenpagdee, M. Diallo, P. van der Heijden, J. Kooiman, R. Mahon and S. Williams (2005). *Interactive Fisheries Governance. A Guide to Better Practice*. Delft: Eburon Publishers.

Belton, B. and D. C. Little (2011). 'Immanent and interventionist inland Asian aquaculture development and its outcomes', *Development Policy Review* 29(4), 459–484.

Belton, B., I. J. M. Asseldonk and S. R. Bush (2016). 'Domestic crop booms, livelihood pathways and nested transitions: Charting the implications of Bangladesh's pangasius boom', *Journal of Agrarian Change* 17(4), 694–714.

Belton, B., S. R. Bush and D. Little (2017a). 'Not just for the wealthy: Rethinking farmed fish consumption in the global South', *Global Food Security* 16, 85–92.

Belton, B., A. Hein, K. Htoo, L. S. Kham, A. S. Phyoe and T. Reardon (2017b). 'The emerging quiet revolution in Myanmar's aquaculture value chain', *Aquaculture* 493, 384–394.

Belton, B., A. Padiyar, G. Ravibabu and K. G. Rao (2017c). 'Boom and bust in Andhra Pradesh: Development and transformation in India's domestic aquaculture value chain', *Aquaculture* 470, 196–206.

Béné, C., B. Hersoug and E. H. Allison (2010). 'Not by rent alone: Analysing the pro-poor functions of small-scale fisheries in developing countries', *Development Policy Review* 28(3), 325–358.

Berkes, F. (1985). 'Fishermen and "the tragedy of the commons"', *Environmental Conservation* 12(3), 199–206.

Berkes, F., T. P. Hughes, R. S. Steneck, J. A. Wilson, D. R. Bellwood, B. Crona, C. Folke, L. Gunderson, H. Leslie and J. Norberg (2006). 'Globalization, roving bandits, and marine resources', *Science* 311(5767), 1557–1558.

Blidariu, F. and A. Grozea (2011). 'Increasing the economical efficiency and sustainability of indoor fish farming by means of aquaponics – Review', *Scientific Papers: Animal Science and Biotechnologies* 44(2), 1–8.

Bondad-Reantaso, M. and R. Subasinghe (2013). 'Enhancing the contribution of small-scale aquaculture to food security, poverty alleviation and socio-economic development', Fisheries and Aquaculture Proceedings No. 31. Rome: FAO.

Bostock, J., B. McAndrew, R. Richards, K. Jauncey, T. Telfer, K. Lorenzen, D. Little, L. Ross, N. Handisyde, I. Gatward and R. Corner (2010). 'Aquaculture: Global status and trends', *Philosophical Transactions of the Royal Society of London B: Biological Sciences* 365(1554), 2897–2912.

Bretherton, C. and J. Vogler (2008). 'The European Union as a sustainable development actor: The case of external fisheries policy', *European Integration* 30(3), 401–417.

Bunting, S. W. (2013). *Principles of Sustainable Aquaculture: Promoting Social, Economic and Environmental Resilience.* Oxford and New York: Routledge.

Bush, S. R. (2017). Understanding the potential of eco-certification in salmon and shrimp aquaculture value chains. *Aquaculture* 493, 376–383.

Bush, S. R. and B. Belton (2012). 'Out of the factory and into the fish pond: Can certification transform Vietnamese pangasius', in G. Spaargaren, P. Oosterveer and A. Loeber (eds) *Food Practices in Transition: Changing Food Consumption, Retail and Production in the Age of Reflexive Modernity*. New York and London: Routledge, 257–290.

Bush, S. R., P. A. van Zwieten, L. Visser, H. van Dijk, R. Bosma, W. F. de Boer and M. Verdegem (2010). 'Scenarios for resilient shrimp aquaculture in tropical coastal areas', *Ecology and Society* 15(2), 15.

Butcher, J. G. (2004). *The Closing of the Frontier: A History of the Marine Fisheries of Southeast Asia,* c.*1850–2000.* Singapore: Institute of Southeast Asian Studies.

Cao, L., R. Naylor, P. Henriksson, D. Leadbitter, M. Metian, M. Troell and W. Zhang (2015). 'China's aquaculture and the world's wild fisheries', *Science* 347(6218), 133–135.

Castello, L., J. P. Viana, G. Watkins, M. Pinedo-Vasquez and V. A. Luzadis (2009). 'Lessons from integrating fishers of arapaima in small-scale fisheries management at the Mamirauá Reserve, Amazon', *Environmental Management* 43(2), 197–209.

Castells, M. (1996). *The Rise of the Network Society. Volume I of The Information Age: Economy, Society and Culture*. Malden, MA and Oxford: Blackwell.

Constance, D. H. and A. Bonanno (1996). *Caught in the Net: Global Tuna Industry, Environmentalists and the State*. Lawrence, KS: University Press of Kansas.

Costa-Pierce, B. A. (2010). 'Sustainable ecological aquaculture systems: The need for a new social contract for aquaculture development', *Marine Technology Society Journal* 44, 88–112.

Cullis-Suzuki, S. and D. Pauly (2010). 'Failing the high seas: A global evaluation of regional fisheries management organizations', *Marine Policy* 34(5), 1036–1042.

Deutsch, L., S. Gräslund, C. Folke, M. Troell, M. Huitric, N. Kautsky and L. Lebel (2007). 'Feeding aquaculture growth through globalization: Exploitation of marine ecosystems for fishmeal', *Global Environmental Change* 17(2), 238–249.

Dierberg, F. and W. Kiattisimkul (1996). 'Issues, impacts, and implications of shrimp aquaculture in Thailand', *Environmental Management* 20(5), 649–666.

Edwards, P. (2015). 'Aquaculture environment interactions: Past, present and likely future trends', *Aquaculture* 447, 2–14.

FAO (2012). *The State of World Fisheries and Aquaculture 2012*. Rome, FAO.

FAO (2014). *The State of World Fisheries and Aquaculture 2014*. Rome, FAO.

FAO (2016a). 'FAOSTAT', FAO, available at http://faostat.fao.org/faostat/ (accessed 29 January 2016).

FAO (2016b). *The State of World Fisheries and Aquaculture 2016*. Rome: FAO.

Finlayson, A. C. (1994). *Fishing for Truth: A Sociological Analysis of Northern Cod Stock Assessments from 1977 to 1990*. St. John's, NL: Institute of Social and Economic Research, Memorial University of Newfoundland St. John's.

Folke, C., N. Kautsky and M. Troell (1994). 'The costs of eutrophication from salmon farming: Implications for policy', *Journal of Environmental Management* 40(2), 173–182.

Folke, C., N. Kautsky, H. Berg, A. Jansson and M. Troell (1998). 'The ecological footprint concept for sustainable seafood production: A review', *Ecological Applications* 8(1) Supplement, 563–571.

Frank, K. T., B. Petrie, J. S. Choi and W. C. Leggett (2005). 'Trophic cascades in a formerly cod-dominated ecosystem', *Science* 308(5728), 1621–1623.

Fréon, P., J. C. Sueiro, F. Iriarte, O. F. M. Evar, Y. Landa, J.-F. Mittaine and M. Bouchon (2014). 'Harvesting for food versus feed: A review of Peruvian fisheries in a global context', *Reviews in Fish Biology and Fisheries* 24(1), 381–398.

Gagern, A. and J. van den Bergh (2013). 'A critical review of fishing agreements with tropical developing countries', *Marine Policy* 38, 375–386.

Hall, D. (2004). 'Explaining the diversity of Southeast Asian shrimp aquaculture', *Journal of Agrarian Change* 4(33), 315–335.

Hall, S. J., A. Delaporte, M. J. Phillips, M. Beveridge and M. O'Keefe (2011). *Blue Frontiers: Managing the Environmental Costs of Aquaculture*. Penang: WorldFish Center.

Hara, M. and F. Njaya (2016). 'Between a rock and a hard place: The need for and challenges to implementation of rights based fisheries management in small-scale fisheries of southern Lake Malawi', *Fisheries Research* 174, 10–18.

Hardin, G. (1968). 'The tragedy of the commons', *Science* 162(3859), 1243–1248.

Hauck, M. (2008). 'Rethinking small-scale fisheries compliance', *Marine Policy* 32(4), 635–642.

Havice, E. (2010). 'The structure of tuna access agreements in the Western and Central Pacific Ocean: Lessons for Vessel Day Scheme planning', *Marine Policy* 34(5), 979–987.

Havice, E. and L. Campling (2010). 'Shifting tides in the Western and Central Pacific Ocean tuna fishery: The political economy of regulation and industry responses', *Global Environmental Politics* 10(1), 89–114.

Havice, E. and L. Campling (2017). 'Where chain governance and environmental governance meet: Interfirm strategies in the canned tuna global value chain', *Economic Geography* 93(3), 292–313.

Hishamunda, N., N. B. Ridler, P. Bueno and W. G. Yap (2009). 'Commercial aquaculture in Southeast Asia: Some policy lessons', *Food Policy* 34(1), 102–107.

Isaacs, M. (2006). 'Small-scale fisheries reform: Expectations, hopes and dreams of "a better life for all"', *Marine Policy* 30(1), 51–59.

Islam, M. S. (2014). *Confronting the Blue Revolution. Industrial Aquaculture and Sustainability in the Global South*. Toronto, ON: University of Toronto Press.

Kaczynski, V. M. and D. L. Fluharty (2002). 'European policies in West Africa: Who benefits from fisheries agreements?', *Marine Policy* 26(2), 75–93.

Kautsky, N., H. Berg, C. Folke, J. Larsson and M. Troell (1997). 'Ecological footprint for assessment of resource use and development limitations in shrimp and tilapia aquaculture', *Aquaculture Research* 28(10), 753–766.

Kautsky, N., P. Rönnbäck, M. Tedengren and M. Troell (2000). 'Ecosystem perspectives on management of disease in shrimp pond farming', *Aquaculture* 191(1), 145–161.

Khalilian, S., R. Froese, A. Proelss and T. Requate (2010). 'Designed for failure: A critique of the Common Fisheries Policy of the European Union', *Marine Policy* 34(6), 1178–1182.

Lebel, L., N. H. Tri, A. Saengnoree, S. Pasong, U. Buatama and L. K. Thoa (2002). 'Industrial transformation and shrimp aquaculture in Thailand and Vietnam: Pathways to ecological, social, and economic sustainability', *Ambio* 31(4), 311–323.

Love, D. C., J. P. Fry, X. Li, E. S. Hill, L. Genello, K. Semmens and R. E. Thompson (2015). 'Commercial aquaponics production and profitability: Findings from an international survey', *Aquaculture* 435, 67–74.

Martins, C. I. M., E. H. Eding, M. C. J. Verdegem, L. T. N. Heinsbroek, O. Schneider, J. P. Blancheton, E. R. d'Orbcastel and J. A. J. Verreth (2010). 'New developments in recirculating aquaculture systems in Europe: A perspective on environmental sustainability', *Aquacultural Engineering* 43(3), 83–93.

Massey, D. and J. Allen (1984). *Geography Matters! A Reader*. Cambridge: Cambridge University Press.

Maurstad, A. (2000). 'To fish or not to fish: Small-scale fishing and changing regulations of the cod fishery in northern Norway', *Human Organization* 59(1), 37–47.

Metian, A. G. T. M. (2009). 'Fishing for feed or fishing for food: Increasing global competition for small pelagic forage fish', *Ambio* 38(6), 294–302.

Munro, G. (2004). 'On the management of shared fish stocks: Critical issues and international initiatives to address them', in A. I. L. Payne, C. M. O'Brien and S. I. Rogers (eds) *Management of Shared Fish Stocks*. Melbourne, Vic.: Blackwell.

Munro, G. (2007). 'Internationally shared fish stocks, the high seas, and property rights in fisheries', *Marine Resource Economics* 22, 425–443.

Myers, R. A. and B. Worm (2003). Rapid worldwide depletion of predatory fish communities', *Nature* 423(6937), 280.

Nayak, P. K. and F. Berkes (2011). 'Commonisation and decommonisation: Understanding the processes of change in the Chilika lagoon, India', *Conservation and Society* 9(2), 132.

Naylor, R. and M. Burke (2005). 'Aquaculture and ocean resources: Raising tigers of the sea', *Annual Review of Environment and Resources* 30, 185–218.

Naylor, R. L., J. Eagle and W. L. Smith (2003). 'Salmon aquaculture in the Pacific Northwest: A global industry with local impacts', *Environment: Science and Policy for Sustainable Development* 45(8), 18–39.

Naylor, R. L., R. Goldburg, J. Primavera, N. Kautsky, M. Beveridge, J. Clay, C. Folke, J. Lubchenko, H. Mooney and M. Torrell (2000). 'Effect of aquaculture on world fish supplies', *Nature* 405, 1017–1024.

Naylor, R. L., R. W. Hardy, D. P. Bureau, A. Chiu, M. Elliott, A. P. Farrell, I. Forster, D. M. Gatlin, R. J. Goldburg, K. Hua and P. D. Nichols (2009). 'Feeding aquaculture in an era of finite resources', *Proceedings of the National Academy of Sciences* 106(36), 15103–15110.

Ngoc, P. T. A., M. P. Meuwissen, L. Cong Tru, R. H. Bosma, J. Verreth and A. O. Lansink (2016). 'Economic feasibility of recirculating aquaculture systems in pangasius farming', *Aquaculture Economics & Management* 20(2), 185–200.

Pauly, D. (1995). 'Anecdotes and the shifting baseline syndrome of fisheries', *Trends in Ecology & Evolution* 10(10), 430.

Pauly, D. (2006). 'Towards consilience in small-scale fisheries research', *Maritime Studies* 4, 7–22.

Pauly, D. (2008). 'Global fisheries: A brief review', *Journal of Biological Research – Thessaloniki* 9, 3–9.

Pauly, D., V. Christensen, J. Dalsgaard, R. Froese and F. Torres (1998). 'Fishing down marine food webs', *Science* 279(5352), 860–863.

Pomeroy, R., K. A. T. Nguyen and H. X. Thong (2009). 'Small-scale marine fisheries policy in Vietnam', *Marine Policy* 33(2), 419–428.

Pomeroy, R., J. Parks, R. Pollnac, T. Campson, E. Genio, C. Marlessy, E. Holle, M. Pido, A. Nissapa and S. Boromthanarat (2007). 'Fish wars: Conflict and collaboration in fisheries management in Southeast Asia', *Marine Policy* 31(6), 645–656.

Puig, P., M. Canals, J. Martín, D. Amblas, G. Lastras, A. Palanques and A. M. Calafat (2012). 'Ploughing the deep sea floor', *Nature* 489(7415), 286.

Roberts, C. (2010). *The Unnatural History of the Sea.* Washington, DC: Island Press.

Salayo, N., L. Garces, M. Pido, K. Viswanathan, R. Pomeroy, M. Ahmed, I. Siason, K. Seng and A. Masae (2008). 'Managing excess capacity in small-scale fisheries: Perspectives from stakeholders in three Southeast Asian countries', *Marine Policy* 32(4), 692–700.

Seto, K., D. Belhabib, J. Mamie, D. Copeland, J. M. Vakily, H. Seilert, A. Baio, S. Harper, D. Zeller and K. Zylich (2017). 'War, fish, and foreign fleets: The marine fisheries catches of Sierra Leone 1950–2015', *Marine Policy* 83, 153–163.

St. Martin, K. (2005). 'Mapping economic diversity in the first world: The case of fisheries', *Environment and Planning A* 37(6), 959–979.

Stobutzki, I. C., G. T. Silvestre, A. Abu Talib, A. Krongprom, M. Supongpan, P. Khemakorn, N. Armada and L. R. Garces (2006). 'Decline of demersal coastal fisheries resources in three developing Asian countries', *Fisheries Research* 78(2–3), 130–142.

Swartz, W., E. Sala, S. Tracey, R. Watson and D. Pauly (2010). 'The spatial expansion and ecological footprint of fisheries (1950 to present)', *PLoS One* 5(12), e15143.

Symes, D. (2012). 'Regionalising the Common Fisheries Policy: Context, content and controversy', *Maritime Studies* 11(1), 6.

Tacon, A. G., M. Metian, G. M. Turchini and S. S. De Silva (2009). 'Responsible aquaculture and trophic level implications to global fish supply', *Reviews in Fisheries Science* 18(1), 94–105.

Troell, M., A. Joyce, T. Chopin, A. Neori, A. H. Buschmann and J.-G. Fang (2009). 'Ecological engineering in aquaculture – Potential for integrated multi-trophic aquaculture (IMTA) in marine offshore systems', *Aquaculture* 297(1), 1–9.

Tsing, A. L. (2011). *Friction: An Ethnography of Global Connection*. Princeton, NJ: Princeton University Press.

Tveteras, R. (2002). 'Industrial agglomeration and production costs in Norwegian salmon aquaculture', *Marine Resource Economics* 17(1), 1–22.

UNDP (2005). *Human Development Report 2005*. Oxford: UN Development Programme.

van Mulekom, L., A. Axelsson, E. P. Batungbacal, D. Baxter, R. Siregar and I. de la Torre (2006). 'Trade and export orientation of fisheries in Southeast Asia: Underpriced export at the expense of domestic food security and local economies', *Ocean & Coastal Management* 49(9–10), 546–561.

Volpe, J. P., J. L. Gee, V. A. Ethier, M. Beck, A. J. Wilson and J. Stoner (2013). 'Global Aquaculture Performance Index (GAPI): The first global environmental assessment of marine fish farming', *Sustainability* 5(9), 3976–3991.

Watson, R., A. Kitchingman, A. Gelchu and D. Pauly (2004). 'Mapping global fisheries: Sharpening our focus', *Fish and Fisheries* 5(2), 168–177.

Watson, R. A., W. W. Cheung, J. A. Anticamara, R. U. Sumaila, D. Zeller and D. Pauly (2013). 'Global marine yield halved as fishing intensity redoubles', *Fish and Fisheries* 14(4), 493–503.

World Bank (2012). 'World Development Indicators 2012', available at data.worldbank.org/data-catalog/world-development-indicators (accessed 1 October 2012).

Yandle, T. and C. M. Dewees (2008). 'Consolidation in an individual transferable quota regime: Lessons from New Zealand, 1986–1999', *Environmental Management* 41(6), 915–928.

Yeeting, A., H. P. Weikard, M. Bailey, V. Ram-Bidesi and S. R. Bush (2017). 'Stabilising cooperation through pragmatic tolerance: The case of the PNA tuna fishery', *Regional Environmental Change* 18(3), 885–897.

Part II

Governing sustainable seafood

4 Public regulation of fisheries and aquaculture

Introduction

The sustainable seafood movement is premised on the perceived inadequacy of states to govern fisheries' and aquaculture sustainability. As seen in various other global food and fibre sectors, globalisation is considered a central driver in undermining the capability of nation-states to cope with production and consumption-related pressures on seafood sustainability. The state sits in a precarious position when it comes to transboundary forms of environmental regulation. Because primary seafood production is enmeshed in global networks of production, unilateral regulation has only a limited effect in achieving sustainability outcomes. Multilevel agreements over trade and resource access are therefore central to state regulation. However, here too the role and effectiveness of the state are brought into question, given that these agreements can be contradictory. As Gulbrandsen (2010) argues, strong multilateral trade agreements can limit the independence of governments to address environmental and social processes of production, while weak agreements often leave the same regulatory 'gap' they seek to overcome.

This chapter discusses whether such analysis of weak states and regulatory gaps is a fair assessment. In doing so we outline the roles of governments acting within their own territories and beyond these through cooperative multilateral arrangements. We start by presenting the role of governments in managing capture fisheries and aquaculture. We then cover governance arrangements that are designed to deal with transnational seafood management, in particular the RFMOs, the WTO and the EU IUU fishing regulation. In the final section we conclude that, despite severe criticisms, states remain essential actors in sustainable seafood governance, including in the context of globalisation, and that therefore the main challenge is not to replace state interventions but to identify appropriate modes of seafood governance involving public as well as private actors.

National governments and the management of seafood provision

National fisheries' policies reflect 'the complex, historically-based interaction of multiple state and non-state interests' (Bush, 2008: 351). Since UNCLOS (1982), and in response to growing awareness about the limits to fishing, national governments have become increasingly engaged in managing fisheries. The creation of EEZs dramatically changed national fisheries' policies. Governments developed and implemented national regulations to manage seafood provision, while also engaging in negotiated decision-making processes involving other governments and non-state private actors (Held et al., 1999; Lipschutz and Fogel, 2002). However, the emergence of EEZs under UNCLOS and the rise of state control over fishery resources have also shifted responsibility for these resources at the local level. As Allison (2001) argues, the increased role of the state in fisheries' management has meant that customary arrangements already in place within communities for fishery management have been gradually eroded, which in turn undermined a local sense of ownership and responsibility for the stewardship of these fishery resources.

Questions of state governance are therefore not only limited to control over resources but also linked to the welfare and well-being of fishing communities. In policy terms, this means that fisheries, and indeed aquaculture development, are seen as programmes for pro-poor economic development (OECD, 2010) and the promotion of food security (HLPE, 2014). Governments may consider the specific interests of the poor when managing fisheries by supporting fishers through targeted revenue policies and sustaining fisheries by shaping rights-based institutions that promote pro-poor growth. As argued by the OECD (2010: 96), 'central to achieving this strategy – and thus to capture the benefits from globalisation – is the development of an appropriate national institutional framework to deliver fisheries governance'. It goes on to argue that for fishery resources to be sustained into the future, developing country governments, with support from donor agencies, need to focus on the development of healthy public governance and institutions.

However, while ownership, control and 'development' have dominated state policy and regulation, sustainability has lagged behind. Effectively promoting sustainable seafood provision often escapes the restricted domain of government interventions. As Ponte et al. (2014) observe, governmental regulatory frameworks have had an important impact on stimulating improvements in seafood production and processing regarding volume and variety, but have had little impact in regulating the quality of production, including sustainability.

Government regulation

Based on the formal principle of national sovereignty, governments have introduced different kinds of regulations to steer fishing practices to become more sustainable. Most governments in developed countries applied a conventional approach to seafood governance (Allison, 2001). Their regulations include a variety of legal and political measures used to steer the behaviour of fishers. But reflecting the discussion in Chapter 3, these regulations often draw a priori divisions between different scales and types of fishers, which can limit their overall effectiveness in reducing pressure on fishery resources.

In general terms, governmental measures are aimed at reducing the exploitation rates of overfished stocks and the pressure on ecosystems (Worm et al., 2006). They can be classified in terms of controlling either *input* or *output* dimensions of fishing activity.

Input measures aim to limit fishing effort, meaning the vessels, gear, fishing days or areas that are used to catch fish. This can mean reducing the number of vessels in a fishery by only allowing licensed boats to operate, or 'buying back' vessels and/or licences. Governments may also limit the efficiency of fishing gear and vessels. Alternatively referred to as 'catchability', reducing gear efficiency includes restrictions on gear design to improve gear selectivity and reduce the bycatch of non-target species (Hilborn and Walters, 1992). Spatial measures are also employed to restrict efficiency, by for instance creating fully protected marine reserves or selectively excluding specific fisheries for a specific period in a particular area.

Output measures aim to regulate the volume of catch relative to the proportion of the available biomass of a fish stock. These measures commonly set a total allowable catch (TAC) based on the fish stock dynamics; this can be divided into quotas that are made transferable. Fishing mortality is commonly regulated by capping the TAC of a particular population of a species (based in principle on sound scientific advice) over a specified time period to reach an agreed-upon level, goal or target such as maximum sustainable yield (Mace, 1994). Fishers can opt for a share of the TAC, an individual (and possibly transferable) quota, and continue fishing as long as the total amount of fish caught does not exceed their share. These output measures can also be linked to input measures by capping total effort – i.e. limiting the total number of vessels or the total number of days at sea. Both catch and effort can then, in principle, be made transferable to other individuals or groups.

These forms of government regulation are all based on the assumption of considerable amounts of available information on both the fish stocks and the fisheries (in terms of numbers of vessels, gears and fishers). They assume that governments have adequate capacity to ensure that any regulation put in place can be monitored, enforced and respond to changes in the status of fish stocks (Walters, 1997; Wilson et al., 2006). They are also in a large part based on the central assumption that property rights can be allocated over both stocks and fishing gear. While such assumptions are generally thought to hold in many developed (OECD) countries, with robust scientific and regulatory information systems and regulatory enforcement, they commonly do not hold in 'information-poor' developing countries (Bentley and Stokes, 2009). However, in practice these measures are questioned across all countries, given issues with overallocation of fishing rights and the exclusion of smaller-scale fishers, including indigenous fishing groups, from accessing stocks in favour of larger industrial fleets.

In summary, government regulation of fisheries is subject to ongoing debate on its effectiveness and indirect impact on non-target species and different groups of fishers (Smith et al., 2010). Central to these debates are not only the effectiveness of such regulation but also the limits of government regulation in the context of global demand for seafood, the predominance of developing countries selling seafood to international markets and the potential for alternative forms of co-management between states and fishing communities, as well as the role of private or non-state actors.

Fisheries' co-management

In response to the exclusion of smaller-scale fishers from fisheries' management and the marginalisation of existing customary or community-based management arrangements, national governments have engaged in decentralised forms of collaborative or 'co'-management (Jentoft, 2000). Co-management implies 'the increased involvement of users of fisheries resources and delegation of decisions to be taken as close to the users as possible' (Nielsen et al., 2004: 154). In practice there is a spectrum of collaboration (Sen and Raakjaer Nielsen, 1996), ranging from hierarchical collaboration led by the state (Bene et al., 2009) to highly devolved community-based management arrangements in which fishing communities are responsible for the full management of fishery resources in a given area, or for a given stock.

Regardless of the degree of collaboration, co-management is seen as a more effective mode of fisheries' management because it internalises incentives for the stewardship of fishery resources. When the state involves itself in these arrangements, it can provide more service-oriented rather than top-down support to fishers. As has been variously demonstrated, even in isolated small-scale developing world fisheries, fishers can understand and change their fishing practices when provided with up-to-date information about their activities (Garaway et al., 2006; Neitzel et al., 2017). In some instances, having such information feeds back to the legitimacy of community-run management in the eyes of both the fishers involved and the state (Pinkerton, 1994; Cinner et al., 2012a, 2012b). Greater transparency in transferring scientific information into fisheries' policy can influence policy effectiveness (Mora et al., 2009; Worm et al., 2009), because the costs of regulation are reduced as policing is needed less. Such arrangements also deliver benefits to governments, which are otherwise unable to monitor the large numbers of small-scale fishers because of a lack of capacity. Without sufficient support from fishers, government measures risk remaining words on paper without much real effect.

A key challenge to co-management of fisheries is how to deal with heterogeneity within local communities, as not all fishers share the same interests and agree unanimously on the measures proposed to address sustainability problems (Paul, 2005; Srinivasan, 2005). In brief, co-management tends to focus on community-based fisheries because these are usually less contentious than so-called middle-scale and/or industrial-scale fisheries – both of which may have far fewer incentives to collaborate with government given that the absence of rules and regulations allows them to (over)exploit fishery resources (Pomeroy et al., 2007). Other contentious issues include how to deal with problems that require solutions at a scale beyond the local community level, and how to reconcile scientific knowledge with local users' knowledge (Nielsen et al., 2004). But perhaps the new frontier of co-management research is the durability of collaboration between communities and the state in the ongoing integration of these fisheries into global markets, as well as their engagement with demands for greater sustainability from international markets.

National governments and the management of aquaculture

Governments engage with regulating aquaculture differently to fisheries because of the specific characteristics of fish farming. Overall, however, the same concerns related to fisheries regarding inclusion and effectiveness of regulation of public resources are at play (see Chapter 3) – but now in relation to water, land and coastal forests. What is striking about the state regulation of aquaculture is that it is often developed in the context of rapid or 'boom' growth (Hall, 2003; Belton et al., 2016). The result is that the expansion of aquaculture, in terms of area, volume and value, has historically exceeded the capacity of governments to regulate the sector. The best documented of these booms and their lack of regulation are shrimp, pangasius and salmon aquaculture (Páez-Osuna et al., 2003; Bush et al., 2009; Young and Matthews, 2011). In all three cases the lack of government regulation was a direct driver of and justification for NGO engagement, and ultimately the rise of instruments such as certification. Overall, however, government regulation of aquaculture remains highly fragmented, with only a handful of countries, including Norway and the US, estimated to have stated aquaculture laws.

Also, as with fisheries, governments have tried to overcome the constraints of poor compliance by working with communities in ways alternative to direct regulation. For example, aquaculture co-management has been introduced in countries such as Sri Lanka and Vietnam to strengthen coordination between the state and producers (Galappaththi and Berkes, 2014). In other countries, including Chile, Thailand and Norway, spatial forms of ecosystem-based management such as clusters and zoning (see Brugère et al., 2018) have been developed.

In most of these examples the driver has been related to disease management rather than to direct environmental concerns – although, as outlined in Chapter 3, disease and environment are linked. The shared interest of governments and communities in collaborating in these arrangements stems from severe declines in production due to disease or lost market access because of failure to comply with export market requirements for food safety. These requirements focus mainly on the use of antibiotics due to human health risks (Reilly and Käferstein, 1997). Import markets like the US and the EU have implemented food safety assurance programmes based on hazard analysis and critical control points (HACCP) to assure the safety of seafood from aquaculture (e.g. Dey et al., 2005; Oosterveer, 2006; Tran et al., 2013). Producers have no choice but to adopt these programmes so as to maintain their export position, although they may be confronted with high costs when implementing them.

The emergence of national standards

Common to government regulation of both fisheries and aquaculture has been the emergence of national production standards, often incorporating a form of farm-level certification and in some cases a product label. The emergence of such approaches in developing countries has been in response to a perceived weak legitimacy of broader (non-farm-level) government regulation, and their desire to (re)gain this legitimacy and strengthen their position in the market.

In Southeast Asia, for instance, the Thai and Vietnamese governments have developed a series of certifications for aquaculture (see Samerwong et al., 2018). In 1997 the Thai government started the Good Aquaculture Practices (GAP) standards, followed by the Thai Quality shrimp labelling scheme in 2004 under the control of the Department of Fisheries. The department prepared a certification scheme for shrimp aquaculture, with the intention of producing shrimp under the government's code of conduct. In 2009 the Thai government developed the GAP-7401 standard in an attempt to upgrade the national certification system and again increase the international market credibility of Thai shrimp aquaculture (Prompoj, 2013). But, as shown by Samerwong et al. (2018), while these standards have been taken up by the industry to varying degrees, their recognition in the international market has remained limited.

In Vietnam the government took a different path to develop a national VietGAP standard by seeking recognition with the European-based retail standard GlobalGAP in 2011. Farmers certified under the VietGAP standard partially conform also with the GlobalGAP standard.

In developed countries, governments such as the US and Iceland have also been developing national standards. Similar to developing countries, they have done so to gain legitimacy in the market; but unlike developing countries they have also done so to compete with private labels such as the MSC. As described by Foley and Hébert (2013), the Alaskan Seafood Marketing Institute, a public–private agency responsible for promoting the 'Alaska brand' of seafood on behalf of all Alaskan fisheries, sought to replace an existing MSC certification with the Alaska Responsible Fisheries Management label – a standard based on the FAO code of conduct and run by the Alaskan Seafood Marketing Institute. Driving this shift to the alternative label was the apparent expectation of Alaskan producers and government that the reputation of their industry in the market would continue with or without MSC certification. In Iceland a similar story has unfolded, with the establishment of the Iceland Responsible Fisheries Management label. How these two labels are likely to correspond to international market demands is taken up in the next chapter.

Transnational management of fisheries and seafood provision

Until the 1950s most fishing took place in territorial seas (generally three nautical miles), but since then fishing fleets have moved further out to sea and distant fishing fleets have become common (Sydnes, 2005). Appropriate governance institutions to manage the pressure on seafood stocks were non-existent, however. In response, some forms of government-based transnational seafood governance are emerging, albeit slowly, through adding an international layer of governance to the existing local and national ones; by a transfer of powers from the local and national to the international-level cross-boundary legal authority, dynamic and complex regulatory arrangements are being created (Hakon Hoel et al., 2005). Box 4.1 illustrates the proliferation of regulations and agreements relevant for contemporary transnational seafood governance and their rapid increase since 1992.

In the second half of this chapter we turn our attention to the role of three transnational institutions that have a substantial impact on state-based seafood regulation: RFMOs, the WTO and the EU IUU fishing regulation.

Box 4.1 Proliferation of international governance in fisheries

Binding agreements

- *1958 Convention on Fishing and the Conservation of the Living Resources of the High Seas:* establishing a general regime for the conservation and rational exploitation of living resources in the high seas.
- *1958 Convention on the Continental Shelf:* establishing a general regime for the continental shelf.
- *1958 Convention on the High Seas:* establishing a general regime for the high seas.
- *1959 Convention concerning Fishermen's Articles of Agreement:* providing minimum standards for entering into and the content of fishermen's articles of agreement.
- *1963 Codex Alimentarius:* protecting consumer health and ensuring fair trade practices in the food industry, promoting coordination of all food standards work.

- *1975 Convention on International Trade in Endangered Species of Wild Fauna and Flora (CITES):* international trade of selected fish species is subject to certain controls; eight species are listed under Appendix 1, for which commercial trade is prohibited, and 28 species under Appendix 2, for which an export permit from the exporting country is required.
- *1982 UNCLOS (UN Convention on the Law of the Sea):* key provisions include the limit of EEZs and confirmation of states' rights to fish the high seas without regulations.
- *1992 Convention on Biological Diversity:* aims to conserve biological diversity, promote the sustainable use of its components and encourage equitable sharing of the benefits arising out of the utilisation of genetic resources.
- *1992 Niue Treaty on Cooperation in Fisheries Surveillance and Law Enforcement:* promotes cooperation in enforcement of the fisheries' laws and regulations of parties and in developing regionally agreed procedures for the conduct of fisheries' surveillance and law enforcement.
- *1993 Agreement to Promote Compliance with International Conservation and Management Measures by Fishing Vessels on the High Seas (FAO Compliance Agreement):* seeks to promote compliance with internationally agreed conservation and management measures by fishing vessels on the high seas, by promoting flag-state responsibility and control over its fishing vessels and the exchange of information on high seas fisheries.
- *1994 WTO Agreement on Technical Barriers to Trade (TBT):* seeks to ensure that technical standards and regulations do not create unnecessary obstacles to trade.
- *1994 WTO Agreement on Application of Sanitary and Phytosanitary (SPS) Measures:* seeks a higher degree of transparency for trading partners and to minimise use of arbitrary trade barriers.
- *1994 WTO Subsidies and Countervailing Measures:* sets the legal framework through the WTO for the application of countervailing duties in trade in fish and fish products.
- *1994 WTO Import Licensing Procedures:* sets the framework for how the WTO deals with import licensing. Import licences provide the means to monitor imports, such as quota fill, origin of imports, flag state, etc.
- *1994 WTO Agreement on Antidumping:* allows governments to act against dumping where there is genuine ('material') injury to the competing domestic industry.
- *1995 UN Fish Stocks Agreement:* limits fishing of fish stocks outside national 200-mile EEZ management sovereignty. The agreement legally binds countries that have ratified it to conserve and manage high-seas fisheries resources. It provides for the establishment of RFMOs.
- *1997 HACCP (Hazard Analysis and Critical Control Points):* a means of ensuring food safety from harvest to consumption.
- *2001 WTO Doha Agenda*: to establish a fair and market-oriented trading system through a programme of fundamental reform including market access, export subsidies and domestic support.

Non-binding agreements

- *1992 UN Agenda 21, Chapter 17:* outlines certain programme areas, including integrated management and sustainable development of coastal areas and

EEZs, sustainable use and conservation of marine living resources of the high seas and strengthening international cooperation and coordination.

- *1992 Declaration of the International Conference on Responsible Fishing:* recommends the FAO to draft the Code of Conduct for Responsible Fisheries.
- *1995 FAO Code of Conduct for Responsible Fisheries:* provides a necessary framework for national and international efforts to ensure sustainable exploitation of aquatic living resources in harmony with the environment by establishing principles and standards applicable to the conservation, management and development of all fisheries, in a non-mandatory manner.
- *1995 Rome Consensus on World Fisheries:* urges governments and international organisations to take prompt action to reduce fishing to sustainable levels and promote sound fisheries' management.
- *1995 Kyoto Declaration on the Sustainable Contribution of Fisheries to Food Security and Plan of Action:* addressing the increasing demand for fishery products.
- *1999 International Plan of Action for the Management of Fishing Capacity:* a voluntary instrument that seeks to achieve worldwide an efficient, equitable and transparent management of fishing capacity.
- *2001 FAO International Plan of Action to Prevent, Deter and Eliminate Illegal, Unreported and Unregulated Fishing:* focuses on all state responsibilities, flag-state responsibilities, coastal state measures, research organisations, RFMOs and the special requirements of developing countries, plus reporting requirements and the role of the FAO in implementing measures to prevent, deter and eliminate IUU fishing.
- *2005 FAO Guidelines on Ecolabelling in the Fisheries Sector:* provides guidance to governments and organisations that already maintain, or are considering establishing, labelling schemes, including the need for reliable, independent auditing, transparency of standard setting and accountability, and the need for standards to be based on good science. They also lay down minimum requirements and criteria for assessing whether a fishery should be certified and an ecolabel awarded, drawing on the FAO's Code of Conduct for Responsible Fisheries.
- *2008 FAO Technical Guidelines on Responsible Fish Trade:* provide practical guidance and advice for policymakers, the fish industry and fisheries' managers on how to implement the provisions of the Code of Conduct for Responsible Fisheries.
- *2008 FAO International Agreement on Protecting Deep-Sea Species and Habitats:* provides guidance on the sustainable use of deep-sea fisheries. Deep-sea fishing should cease in areas where significant adverse impact on vulnerable ecosystems is taking place. Conservation and management measures should be put in place to achieve the sustainable use and long-term conservation of deep-sea fish stocks. The guidelines also outline steps for improving information on the location and status of vulnerable deep-sea ecosystems.
- *2015 UN Sustainable Development Goals:* set a general agenda for global development. SDG 14 is specifically on seafood: 'Conserve and sustainably use the oceans, seas and marine resources for sustainable development.'

Source: OECD (2010)

Regional fisheries' management organisations

Rules for the exploitation of transboundary fish stocks are set by RFMOs, with authority mandated through the 1982 UNCLOS. RFMOs require coastal and fishing states to cooperate through international agreements to ensure the conservation and promotion of optimum utilisation of highly migratory species within and beyond the EEZs. Under international law (UNCLOS), RFMOs are expected to establish appropriate and precautionary management measures. In practice, RFMOs collect fishery statistics, assess the state of resources, analyse management options, provide scientific advice for the management of fisheries and design policy measures.[1] From an intergovernmental perspective, they are considered to be the 'the primary organisational mechanism through which states work together to ensure the long-term sustainability of shared fishery resources' (FAO, 2014: 81).

The coverage of RFMOs is indeed impressive. There are currently 16 RFMOs, of which five (covering approximately 91 per cent of the world's oceans) are tasked with the management of tuna. Eight can be considered as 'general' RFMOs responsible for all organisms within their areas, and the remaining three are specialised RFMOs with more specific areas of responsibility around salmon, pollock and marine mammal management (Ásmundsson, 2016). All states that have a stake in a particular region or fishery may take part in an RFMO. For instance, the EU is participating in all 16 RFMOs,[2] including the six tuna-based ones and the region-based ones in the Indian and the Pacific Oceans.

Despite their achievements as international platforms for the management of fishing resources, the RFMOs' effectiveness in setting and enforcing effective effort limits and rules for harvesting fisheries and other marine organisms remains restricted, even when measured against their own goals. Many of these perceived 'failures' result from the difficulty of aligning the political and economic interests of member states (Grafton et al., 2006; Squires et al., 2016). For example, one area exemplifying the limitations faced by the RFMO process has been the establishment of harvest strategies that include harvest control rules (HCRs) and reference points to promote a precautionary approach to fisheries' management as required under UNCLOS and the FAO Code of Conduct for Responsible Fisheries (Hilborn et al., 2005).

Harvest strategies are considered to represent a best practice approach to fisheries' management. They are proactive, adaptive and provide a framework for taking the best available information about a stock or fishery and applying an evidence-based and risk-based approach to setting biological targets and limit reference points (harvest levels). The challenge in setting these rules and levels is that they should be developed in the management planning stage with the involvement of all stakeholders (ibid.; Hampton et al., 2012). However, the complex nature of many fisheries, including multiple species and gear types, contributes to the difficulties in member states cooperating over sustainable management. As Aranda et al. (2012) argue, the adoption (or not) of HCRs and reference points by RFMOs is a function of the (mis)alignment of political and economic goals of participating states given their vested interests in different species, the fishing gears used to catch them and the protection of often long-standing access arrangements (see also Bailey et al., 2013; WCPFC, 2014).

Despite their difficulties, RFMOs remain a critical building block for the management of transboundary stocks – a position reiterated in different international declarations. For instance, the 2013 UN General Assembly resolution on 'Sustainable fisheries' (UN, 2013) underlined national governments' obligation (since the 1982 UNCLOS) to cooperate through RFMOs in conserving and managing living marine resources. However, perhaps most illustrative of the ongoing relevance of RFMOs came out of the 2010 controversy over Atlantic bluefin tuna, a species that has been subject to persistent overfishing (Webster, 2011). During the meeting of the Convention on International Trade in Endangered Species (CITES) in 2010 the member countries decided not to list bluefin tuna – which would have effectively ended fishing completely. However, they did so against the demands of many conservationists, who feared the complete disappearance of this migratory species, but to the satisfaction of Japan, which lobbied against such a ban. The participants considered the International Commission for the Conservation of Atlantic Tunas, which has been the subject of intense scrutiny in the past, as better placed to manage the stocks of bluefin tuna. The reason given was that, as an RFMO, the organisation is best positioned to bring relevant stakeholders together to continue working towards binding HCRs and limited reference points.

The World Trade Organization and seafood trade

The WTO is a multilateral institution and also influences transnational seafood governance (Shaw, 2000), because although the WTO deals with regulating international trade, its activities have direct impacts on seafood policies, particularly on government subsidies and trade regulations. It does so under its overall goal of promoting international trade in general: although seafood is not specifially covered by the WTO, seafood trade-related issues are included in several of the organisation's activities.

The WTO is the central multilateral institution in regulating international trade, subsidies, food safety requirements and many other issues related to importing and exporting goods and services. The organisation was established in 1995 as the successor to the General Agreement on Trade and Tariffs (GATT), which was founded in 1945 to promote international trade and harmonise relevant national regulations. The WTO is a membership organisation which today has some 160 members – all national governments and the EU. The organisation's stated objective is to facilitate international free and fair trade in support of global welfare by removing all 'unnecessary and unjustified' barriers to trade, such as import levies and quantitative limitations to import, as well as subsidies that limit an individual country's potential to make use of its comparative advantages in international trade. While the WTO recognises individual countries' sovereignty in determining decision-making on commerce and trade, national policy should not unjustifiably complicate international trade. At the same time, respecting nation-states' sovereignty means that governments are not allowed to interfere with the internal and domestic policies of other member states.

Out of fear of protectionism, the WTO allows governments to interfere with international trade only on the basis of product-related characteristics, and to intervene with 'process and production methods' only when these directly relate to the properties of the traded product itself; WTO members may not intervene in international trade on the basis of non-product-related process and production methods. So production methods must have a verifiable impact on the final product before they may be subjected to regulation, otherwise the way in which a product is produced cannot be included in import requirements. For example, whether or not a fish is raised or captured sustainably cannot be a condition for international trade (Oosterveer, 2008) according to the WTO.

Only under very exceptional conditions may the practices involved in capturing and processing seafood be included in trade regulation. Perhaps the most prominent case of discussing this and applying the WTO rules on seafood production practices was the dolphin–tuna case brought before the GATT Dispute Panel (see Box 4.2).

The WTO approach to global seafood governance is based on securing and facilitating global trade flows rather than guaranteeing sustainability in fisheries or on aquaculture farms. As confirmed in the latest decisions on the 'dolphin-safe' tuna case, the WTO does not prohibit governments from intervening for reasons of sustainability provided such measures are not applied for protectionist purposes. As reflected in other cases (see DeSombre and Barkin, 2002), the result is that within the WTO framework a government is allowed to take measures only to protect an animal species that is under threat of extinction.

Seafood trade is also subject to other regulations within the WTO, notably two separate covenants: the Sanitary and Phytosanitary (SPS) Measures agreement and the Technical Barriers to Trade (TBT) agreement (Appleton, 1999; Echols, 2001; Motaal, 1999). These agreements deal with food safety and standards in international trade. Preventing food risks, including from seafood, to protect human health is an acceptable goal according to the WTO. Food safety measures may be imposed upon imported seafood products provided they are based on sound scientific evidence and address objectively verifiable product characteristics.

The SPS agreement is part of the Agreement on Agriculture that was concluded within the WTO in 1995, and intends to distinguish between the justified use of trade-restrictive measures for human health and their unjustified use for reasons of protecting national interests. Some developing countries

Box 4.2 The dolphin–tuna dispute

In 1988 the US environmental organisation Earth Island Institute (EII) filed a legal complaint against the US federal government, arguing that the US had a legal obligation to curtail the incidental killing of dolphins by Mexican tuna fishermen in the tropical Eastern Pacific Ocean. The EII argued that this tuna should not be allowed to be imported into the US.

The federal court agreed, and in 1991 Mexican tuna imports were to be banned from the US market. In response Mexico argued that its right to sell tuna in the US had been violated unjustifiably, and asked GATT (the precursor of the WTO) to solve the matter. It argued the US was not respecting its obligation not to restrict imports on the basis of production methods. The panel concluded that the US was indeed in violation of GATT, but Mexico decided not to pursue the case further and the panel report was never formally adopted. Instead, Mexico and the US decided to have bilateral consultations to reach an agreement outside GATT.

In 2003 the US changed the law and developed a 'dolphin-safe' label for tuna that was captured using particular non-harmful fishing methods. Today many retailers around the world consider the 'dolphin-safe' label a standard requirement for all canned tuna. However, as Mexican fishermen applied different fishing methods to prevent catching dolphins than those prescribed in the US regulation, they could not be labelled as 'dolphin-safe'.

The Mexican fishermen then argued that innovations in fishing techniques had resulted in lower mortality of dolphins, leading the Mexican government to bring a new case to the WTO in 2008. The WTO dispute settlement body concluded in 2011 that the 'US policy is more restrictive than necessary, so not acceptable'. This led to a new round when both Mexico and the US appealed in the WTO in 2012, with the decision upheld by the highest WTO court. In reaction the US revised its regulation, but not to the satisfaction of Mexico. This led to a further WTO panel in 2013 to verify whether the US was actually implementing the WTO ruling. This panel decided, in April 2015, that the US government had not made enough revisions in its labelling measure to allow Mexican tuna products fair competition.

In June 2015 the US decided to appeal this decision again because it considered its regulations fair for Mexican tuna fishers, while changing them would result in passing 'inaccurate information ... on to consumers'. In return Mexico appealed that the judgment violated the WTO's non-discrimination requirement. When the panel and the appellate panel ruled in 2015 that the US needed to revise its policy further, the government announced a set of changes. And again these changes are subject to another round of compliance panel proceedings, still ongoing in 2017.

Sources: Miller and Bush (2015); Vogel (2000); DeSombre and Barkin (2002); ICTSD (2011)

fear that non-tariff barriers may constitute a challenge for international seafood trade because seafood is perishable and sensitive for food safety risks, and therefore cannot handle delay in transport easily. On the other hand, developed countries may be pressurised by domestic retailers and consumers to impose complex quality requirements, as well as health and safety conditions, for seafood in exporting countries. There seems to be a thin line between 'legitimate' measures taken to safeguard consumers and the environment and 'illegitimate' measures taken to protect national producers. However, so far the SPS agreement has not played an important role in international seafood trade, contrary to several other food products (such as hormone-treated beef; see Kastner and Pawsey, 2002).

The TBT agreement was adopted in 1979 within GATT, to warrant that national product standards and technical regulations would not create unnecessary barriers to international trade, and the agreement is maintained within the WTO. Its different articles cover technical regulations (mandatory provisions), standards (voluntary provisions) and general provisions. The core principle of the TBT agreement is that WTO members have agreed to treat imported products no less favourably than similar (or 'like') products that are produced domestically. Member states are advised to use internationally agreed-upon technical standards and regulations whenever they exist, but may apply their own national standards and regulations provided they are based on sufficient scientific evidence and fulfil the criteria of objectivity, non-discrimination and proportionality. Similarly to the SPS agreement, the TBT agreement has hardly been involved in seafood issues, except for a discussion between the EU and Peru on the definition of 'sardines' (WTO, 2003). Nevertheless, the existence of the two agreements positively influences international trade regulations, as governments fear being accused of protectionism when introducing new regulations on safety and sustainability of seafood.

Seafood labels, which we turn to in the next chapter, are a contentious issue and might be considered by the WTO as a subject for regulation. Developing countries fear that private certification requirements may become a barrier to entry into the lucrative fish markets of Western countries. However, so far this has not been addressed by the WTO because the organisation only deals with mandatory standards and regulations that are introduced by member states' governments and only acts on complaints filed by its members. Sustainable seafood labels are voluntary, initiated and managed by NGOs and private companies, mostly without active government involvement. Such schemes seem not challengeable at the WTO because they are not mandatory, not included in official government regulations and not directly supported by governments. Although the issue continues to be debated (Maciel, 2015), so far it has not led to the actual involvement of the WTO.

The World Trade Organization and fisheries' subsidies

Seafood trade is explicitly included in two ongoing Doha Round negotiations within the WTO: first those on 'non-agricultural market access', and second those on the discipline of the Agreement on Subsidies and Countervailing Measures, discussed within the Negotiating Group on Rules (Potts and Haward, 2007). The primary objective of the non-agricultural market access negotiations is to reduce import tariffs to facilitate trade, but as they are already low for seafood, this issue no longer seems very relevant. Further abolishing fishery tariffs might even be problematic for some African countries, because they currently have preferential access to the EU market and this may erode if other countries can compete on the same basis.

The negotiations taking place within the Negotiating Group on Rules for the Agreement on Subsidies and Countervailing Measures are of much greater relevance. This group is charged with 'reducing or eliminating fisheries subsidies' that enhance capacity, a goal that was included in the Trade and Environment section of the Doha Declaration (WTO, 2001) and supported by SDG 14: 'Conserve and sustainably use the oceans, seas and marine resources for sustainable development.'[3] WTO members have agreed to strengthen their discipline regarding fisheries subsidies, 'with a view to enhancing the mutual supportiveness of trade and environment'. In fisheries direct subsidies are substantial, and many scientists and politicians consider these to have distorting effects on production costs and therefore on competitive positions in international trade.[4] Worldwide annual government spending on fisheries' subsidies in 2009 were estimated to be around US$35 billion, of which some US$20 billion (Sumaila et al., 2013) were considered 'capacity enhancing', particularly through fuel subsidies (Binet, 2007; Sumaila et al., 2010a) and boat construction and renovation programmes.[5] Fuel subsidies represented 23 per cent of these subsidies, while boat construction and renovation were responsible for 11 per cent (Sumaila et al., 2010b).[6] Of all countries, Japan, China and the EU were estimated to be the largest subsidisers, spending about US$4.6 billion, US$4.1 billion and US$2.7 billion per year respectively. Subsidies are considered responsible for environmental degradation because they encourage further growth in the capacity of fishing fleets and thus overfishing (FAO, 2000; Potts and Haward, 2007; Markus, 2010; Sumaila et al., 2010b).

To strengthen the discipline on subsidies in the fisheries' sector, different submissions have been discussed. One proposal was introduced by the 'Friends of Fish' group, and aims to eliminate subsidies for the construction of new fishing vessels and operating costs of such vessels. The Friends of Fish group is composed of countries that intend to promote trade (such as New Zealand, Chile, Iceland and the US) and backed by environmental NGOs, notably the World Wide Fund for Nature (WWF) and Friends of the Earth International (Potts and Haward, 2007). Their goal is to make an exception for subsidies in developing countries and for some measures that are identified as acceptable after negotiations.[7] Their proposal is claimed to be beneficial for trade, but also for protecting the marine environment and promoting sustainable development. Tough seafood subsidies disciplines are considered an essential complement to strong fisheries' management programmes to ensure that wild fish stocks remain in existence for the future. This proposal received important scientific support from fisheries experts worldwide,[8] but is opposed by countries like Japan, Korea and Taiwan, which argue that there is no need for a general ban on fisheries' subsidies, except for those that are problematic from an international trade point of view. These countries argue that subsidies cannot be simply categorised as trade distorting or threatening the environment, because the negative impact on fishing capacity has to be demonstrated much more concretely before they can be prohibited (Oosterveer, 2008). The presence of these conflicting views and the absence of a compromise stalled the negotiations on fisheries' subsidies within the WTO, because the organisation requires consensus among its members for decision-making.[9] SDG target 14.6,[10] however, re-energised the discussion because it explicitly calls for ending harmful fisheries' subsidies by 2020.

During these negotiations, developing countries demanded specific attention to their situation. Countries with limited fishing capacity argued that their small developing economies with their artisanal and small-scale fishing practices have such a minimal impact on global problems of overfishing and overcapacity that wide-ranging prohibitions on subsidies to protect the environment would be unduly punitive for them. There are, however, considerable differences between developing countries: some indeed have a mostly artisanal fishing industry, others have considerable fisheries' resources but little capacity to use them, such as small-island developing states, and some countries, such as Bangladesh, are important seafood exporters. In the negotiations on fisheries' subsidies, developing countries insist on special and differential treatment and exemption from access fees.[11] Particularly the EU, the US and Japan have established fisheries' access agreements with governments in Africa (Alder and Sumaila, 2004), Pacific island states and East Asia. The OECD (2010) reports that some 4.5 million MT, or some 5 per cent of total marine capture, comes from distant-water fisheries covered by fisheries' access agreements. The compensation may differ considerably, but is estimated to represent on average 6 per cent of the harvest (Mbithi Mwikya, 2006). The income from these agreements may be substantial for some countries, as illustrated by some small Pacific island states where 50 per cent of their GDP comes from these payments (ibid.). Still, whether these payments provide full compensation for the resources used is highly questionable, and the actual payments even seem to decrease over the years (Le Manach et al., 2013). There are no indications that the small fishing communities in the countries concerned are profiting from these agreements (Alder and Sumaila, 2004). However, it seems that by now the expansion of these agreements is ending (OECD, 2010), as most coastal states have developed sufficient capacities to exploit their EEZs themselves.[12]

Overall, the involvement of the WTO has had direct and indirect consequences for global seafood provision: directly through the implementation of measures promoting international trade, limiting the domains in which governments are allowed to intervene and the debates on abolishing fisheries' subsidies; and indirectly by limiting the possibilities for the inclusion of environmental considerations in trade regulations.

Illegal, unreported and unregulated fishing

IUU fishing is seen as an additional threat to fish stocks that are already under pressure (Gallic and Cox, 2006; Agnew et al., 2009; FAO, 2014). *Illegal* refers to fishing activities that violate the applicable national and international laws; *unreported* refers to fishing activities that do not meet the requirements for an area or species; and *unregulated* refers to fishing activities by vessels flying the flag of a state not involved in the relevant RFMO or no flag at all (FAO, 2001).

IUU fishing activities include

> incursions into EEZs and inshore areas by foreign fishing vessels … unauthorized fishing in restricted areas, incomplete catch and effort reporting by industrial fleets, fishing by unauthorized vessels, lack of compliance by fishers with the terms of their authorization, underreporting of catches, use of banned gears, fishing with explosives and poisons and the use of dams for the illegal netting of fish in inland fisheries.
>
> (Garcia and Doulman, 2005: 175)

It is domestic-flagged rather than foreign-flagged vessels that are responsible for IUU fishing (Sumaila et al., 2014). As Schmidt (2004: 4) argues, IUU fishing may be a spillover effect from the increasingly stringent regulations on fishing given that high-seas IUU fishing 'doubled over the past twenty years' during a time 'when domestic fishing activities have been increasingly restricted'. IUU fishing is particularly attractive for species that have reduced stocks and are strictly regulated because there is high market demand and thus high profits as well, as illustrated in the case of the Patagonian toothfish.[13] The total value of illegal and unreported fishing losses worldwide are estimated to range between US$10 billion and US$23.5 billion annually, representing 11–26 million MT of seafood (Agnew et al., 2009). In the case of the US, IUU fishing was estimated to be responsible for 20–32 per cent of the wild-caught marine imports entering the country in 2011, representing a value of US$1.3–2.1 billion (Pramod et al., 2014).

IUU fishing has negative consequences for fish stocks and threatens the environment, but it also undermines labour standards, harms markets (OECD, 2015) and reduces prospects for effective fisheries' management. These IUU fishing practices are combatted by different strategies, such as increased international governmental collaboration, strengthening the capacity of weak enforcement agencies and more market transparency (Lack, 2007; Miller and

Sumaila, 2016; Schmidt, 2017). Effectively addressing IUU fishing is seen as an essential goal for national governments, RFMOs and the FAO (Garcia and Doulman, 2005).[14] Countries develop their policies on the basis of their specific situations (Schmidt, 2004), using the 2001 FAO international plan of action on IUU fishing (FAO, 2001) as a toolbox (Leroy et al., 2016). RFMOs may support the development and implementation of these national policies (see for instance Lack, 2007). The (2009) FAO Agreement on Port State Measures to Prevent, Deter and Eliminate Illegal, Unreported and Unregulated Fishing is expected to make an additional contribution because it blocks the landing and trade of fish caught by IUU fishing (Camilleri and Lem, 2014; FAO, 2014). The FAO has also developed a database on fishing vessels, fishing on the high seas and its authority status on fishing in the high seas which facilitates monitoring of fishing vessels (ibid.).

A recent policy innovation from the EU, a ban on discard, is aimed at further increasing transparency in fisheries and reducing illegal practices. Traditionally, a part of the captured fish is discarded by fishers for commercial or regulatory reasons. Some species have little demand in the market and are therefore commercially unattractive, while for other species fishers may not have permission or the catch is below the minimum landing size. These fish are thrown overboard, but often they do not survive. Kelleher (2005) reports that European discards are about 1.3 million MT per annum or 13 per cent of the total catches, which has substantial effects on fish stocks. Following the example of some other countries (Condie et al., 2014), the EU decided to install a discard ban in several annual steps according to different species, starting in 2015 with the pelagic species. The measure is intended to encourage vessels to apply more selective fishing techniques and acquire a quota mix that reflects their actual fishing patterns. If vessels have unintended catches, they should identify and release alive those species which have a known high survival rate, land the rest of their catch and sell undersized fish either as fishmeal without financial gains or commercially on the market but under quota regulations. This ban forces fishers to maintain a complete administration of their fishing efforts and their catches while being monitored through electronic systems (ibid.). Some doubt the effectiveness of these measures because the direct impacts are probably small and only work in combination with additional measures, depending on the species concerned (ibid.; Borges, 2015; Sardà et al., 2015). Nevertheless, by increasing transparency on seafood catches and trade, the discard ban and the IUU fishing regulations may contribute to reducing pressure on fishing resources under threat.

Discussion and conclusions

States are confronted with difficult challenges when trying to address sustainable seafood provision. In short, this is because international trade is increasing and sustainability problems are expanding and intensifying at the same time as their capacities to respond are declining.

National governments are expected to steer seafood provisioning towards more sustainability through interventions such as installing EEZs and imposing quotas, Marine Protected Areas, IUU fishing regulations and discard bans, and refrain from capacity-enhancing measures such as subsidies for boat construction or fuel. However, in practice very few governments have developed and successfully implemented such integrated fisheries' policies. As outlined in this chapter, efforts at a transnational level have been made by governments engaging in the management of straddling stocks through RFMOs, steering seafood trade through the WTO, discussing interventions through the FAO and applying multilateral agreements such as CITES. Despite these efforts, however, governmental interventions have not yet resulted in effective and sustainable fisheries' management, as the continued pressure on fish stocks worldwide illustrates.

Indeed, it is precisely these observations that support arguments for replacing the state by the introduction of market-based or civil-society-initiated seafood supply-chain governance. Concluding that the problems in global seafood governance are another illustration of state failure in environmental governance and that private arrangements should replace the state is not necessarily fully correct (nor new, see Dryzek, 1987; Jänicke, 1990; Eckersley, 1995). States cannot be ignored when managing fisheries and seafood trade because their role is essential in two respects. First, only governments have the necessary authority and legitimacy to intervene in seafood management within national territories. Second, governments may, in principle, counterbalance the dominance of economic interests and protect social and environmental concerns.

The primary challenge is therefore not to develop alternatives with the intention of replacing state interventions, but to develop adequate governance arrangements wherein states can engage effectively. Some optimistically argue for the creation of a global fisheries' management organisation to strengthen the role of governments, 'in the form of an institution designed to address the many problems that face global fisheries regulation' (Barkin and DeSombre, 2013: 156). However, considering the challenges faced by existing multilateral organisations, it is quite unlikely that such an organisation will be effective very soon. Better prospects may be found in balancing public and private interventions in sustainable seafood governance. With new kinds of partnerships between the state and private actors (including local communities) (Allison, 2001; May, 2008), it is increasingly clear that, in the words of Mansfield (2007: 1932), 'market-oriented regulation is not supplanting but rather supplementing an existing regulatory structure'.

In the following chapter we turn to the emergence of so-called market-based approaches to seafood governance. We build on the observations in this chapter to explore the extent to which market partners require active government involvement in governing seafood in globalising value chains.

Notes

1 See www.fao.org/fishery/topic/16800/en (accessed 8 August 2017).

2 See http://ec.europa.eu/fisheries/cfp/international/rfmo/index_en.htm (accessed 8 August 2017).

3 Calculating fishery subsidies (broadly defined as 'financial contribution by the public sector that provides private benefits to the fisheries sector' in WorldBank and FAO, 2009) is fraught with difficulties. However, in general there is consensus that subsidies are substantial and two categories exist: subsidies to support access to fish stocks and subsidies for seafood trade. Further agreement is lacking because definitions are disputed, data are lacking and the impact subsidies have is being debated.

4 See the WTO Ministerial Declaration after the Hong Kong meeting, 18 December 2005, Annex D, article 9.

5 Fisheries' subsidies with a direct impact on fishing capacity per year were estimated by the World Bank and FAO (2009) to be US$10 billion in 2000, of which 63.5 per cent was for fuel and 18.9 per cent was for vessel construction, renewal and modernisation.

6 In developing countries with mostly rather small-scale production units, subsidies are not a common feature, while developed countries with a large-scale fishery sector may have elaborate subsidy arrangements influencing capacity and the realised fishing efforts (FAO, 2007).

7 See www.wto.org/english/tratop_e/rulesneg_e/fish_e/fish_intro_e.htm (accessed 12 June 2014).

8 These scientists write in their letter dated 24 May 2007:

> The world's ocean ecosystems are at a tipping point, and overfishing represents one of the greatest threats to their productivity. Yet despite the precarious state of the oceans, many governments continue to provide significant subsidies to their fishing sector. Fisheries subsidies produce such strong economic incentives to overfish that reducing them is one of the most significant actions that can be taken to combat global overfishing. The WTO negotiations represent the best opportunity to control worldwide subsidies to the fishing sector.
>
> See http://oceana.org/fileadmin/oceana/uploads/reports/Scientists_Letter_FINAL_5_24_07.pdf (accessed 30 May 2007)

9 This stalemate in negotiations on fisheries subsidies within the WTO coincides with a general collapse in the WTO Doha Round (Pritchard, 2009).

10

> By 2020, prohibit certain forms of fisheries subsidies which contribute to overcapacity and overfishing, and eliminate subsidies that contribute to IUU fishing, and refrain from introducing new such subsidies, recognizing that appropriate and effective special and differential treatment for developing and least developed countries should be an integral part of the WTO fisheries subsidies negotiation.
>
> See http://indicators.report/targets/14–6/ (accessed 12 September 2017)

11 Fisheries' access fees, or fishing rights, are bilateral contractual agreements, often between developed countries and developing coastal states that lack the capacity to capitalise on their own national fisheries' resources. Through these agreements, one nation's fishing fleet is allowed access another country's EEZ in return for compensation (financial or otherwise).

12 Le Manach et al. (2013) argue differently, and claim that because the payments are low they have not led to domestic fishing capacity development.

13 See http://thefishproject.weebly.com/illegal-unreported-unregulated-iuu-fishing1.html (accessed 10 November 2014).

14 See Österblom and Bodin (2012) for a case on cooperation in reducing illegal fishing in the Pacific Ocean. Also the EU is banning seafood imports from countries that have taken insufficient measures to comply with their international obligation to prevent illegal fishing, such as Sri Lanka and Thailand. See ICTSD (2014, 2015).

References

Agnew, D. J., J. Pearce, G. Pramod, T. Peatman, R. Watson, J. R. Beddington and T. J. Pitcher (2009). 'Estimating the worldwide extent of illegal fishing', *PLoS One* 4(2), e4570.

Alder, J. and U. R. Sumaila (2004). 'Western Africa: A fish basket of Europe past and present', *Journal of Environment & Development* 13(2), 156–178.

Allison, E. H. (2001). 'Big laws, small catches: Global ocean governance and the fisheries crisis', *Journal of International Development* 13(7), 933–950.

Appleton, A. E. (1999). 'Environmental labelling schemes: WTO law and developing countries implications', in G. P. Sampson and W. B. Chambers (eds) *Trade, Environment, and the Millennium*. Tokyo: United Nations University Press, 195–222.

Aranda, M., H. Murua and P. de Bruyn (2012). 'Managing fishing capacity in tuna regional fisheries management organisations (RFMOs): Development and state of the art', *Marine Policy* 36(5), 985–992.

Ásmundsson, S. (2016). 'Regional fisheries management organisations (RFMOs): Who are they, what is their geographic coverage on the high seas and which ones should be considered as general RFMOs, tuna RFMOs and specialised RFMOs?', Convention on Biodiversity, available at www.cbd.int/doc/meetings/mar/soiom-2016-01/other/soiom-2016-01-fao-19-en.pdf (accessed 17 October 2017).

Bailey, M., U. Sumaila and S. Martell (2013). 'Can cooperative management of tuna fisheries in the Western Pacific solve the growth overfishing problem?', *Strategic Behavior and the Environment* 3, 31–66.

Barkin, J. S. and E. R. DeSombre (2013). *Saving Global Fisheries: Reducing Fishing Capacity to Promote Sustainability*. Cambridge, MA: MIT Press.

Belton, B., I. J. M. Asseldonk and S. R. Bush (2016). 'Domestic crop booms, livelihood pathways and nested transitions: Charting the implications of Bangladesh's pangasius boom', *Journal of Agrarian Change* 17(4), 694–714.

Bene, C., E. K. Abban, P. Van Zwieten, H. R. Dankwa, R. Brummett, J. K. Ofori, N. Obirih-Opareh and J. Kolding (2009). *Engaging Local Communities in Aquatic Resources Research and Activities: A Technical Manual*. Penang: WorldFish Center.

Bentley, N. and K. Stokes (2009). 'Contrasting paradigms for fisheries management decision making: How well do they serve data-poor fisheries?', *Marine and Coastal Fisheries: Dynamics, Management, and Ecosystem Science* 1(1), 391–401.

Binet, T. (2007). *Fuelling the Threat for Sustainable Fisheries in Europe*. Brussels: WWF European Policy Office.

Borges, L. (2015). 'The evolution of a discard policy in Europe', *Fish and Fisheries* 16(3), 534–540.

Brugère, C., J. Aguilar-Manjarrez, M. C. Beveridge and D. Soto (2018). 'The ecosystem approach to aquaculture 10 years on – A critical review and consideration of its future role in blue growth', *Reviews in Aquaculture*, available at https://onlinelibrary.wiley.com/doi/abs/10.1111/raq.12242 (accessed 20 November 2018).

Bush, S. R. (2008). 'Contextualising fisheries policy in the lower Mekong Basin', *Journal of Southeast Asian Studies* 39(3), 329–353.

Bush, S. R., N. T. Khiem and L. X. Sinh (2009). 'Governing pangasius for sustainable rural livelihoods and environmental performance: A review', *Aquaculture Economics and Management* 13(4), 271–293.

Camilleri, M. and A. Lem (2014). 'Tackling illegal, unreported, and unregulated fishing: New approaches', *BioRes* 8(6), 4–7.

Cinner, J. E., T. Daw, T. McClanahan, N. Muthiga, C. Abunge, S. Hamed, B. Mwaka, A. Rabearisoa, A. Wamukota and E. Fisher (2012a). 'Transitions toward co-management: The process of marine resource management devolution in three East African countries', *Global Environmental Change* 22(3), 651–658.

Cinner, J. E., T. R. McClanahan, M. A. MacNeil, N. A. Graham, T. M. Daw, A. Mukminin, D. A. Feary, A. L. Rabearisoa, A. Wamukota and N. Jiddawi (2012b). 'Comanagement of coral reef social-ecological systems', *Proceedings of the National Academy of Sciences* 109(14), 5219–5222.

Condie, H. M., A. Grant and T. L. Catchpole (2014). 'Incentivising selective fishing under a policy to ban discards: Lessons from European and global fisheries', *Marine Policy* 45(C), 287–292.

DeSombre, E. R. and J. S. Barkin (2002). 'Turtles and trade: The WTO's acceptance of environmental trade restrictions', *Global Environmental Politics* 2(1), 12–18.

Dey, M. M., M. A. Rab, K. M. Jahan, A. Nisapa, A. Kumar and M. Ahmed (2005). 'Food safety standards and regulatory measures: Implications for selected fish exporting Asian countries', *Aquaculture Economics & Management* 9(1–2), 217–236.

Dryzek, J. S. (1987). *Rational Ecology. Environment and Political Economy.* Oxford: Blackwell.

Echols, M. A. (2001). *Food Safety and the WTO. The Interplay of Culture, Science and Technology.* The Hague: Kluwer Law International.

Eckersley, R. (ed.) (1995). *Markets, the State, and the Environment: Towards Integration.* Melbourne, Vic.: Macmillan Education Australia.

FAO (2000). 'International trade in fishery products and the new global trading environment', in *Multilateral Trade Negotiations on Agriculture: A Resource Manual.* Rome: FAO.

FAO (2001). *International Plan of Action to Prevent, Deter and Eliminate Illegal, Unreported and Unregulated Fishing.* Rome: FAO.

FAO (2007). *The State of the World Fisheries and Aquaculture 2006.* Rome: FAO.

FAO (2014). *The State of World Fisheries and Aquaculture 2014.* Rome: FAO.

Foley, P. and K. Hébert (2013). 'Alternative regimes of transnational environmental certification: Governance, marketization, and place in Alaska's salmon fisheries', *Environment and Planning A* 45(11), 2734–2751.

Galappaththi, E. K. and F. Berkes (2014). 'Institutions for managing common-pool resources: The case of community-based shrimp aquaculture in northwestern Sri Lanka', *Maritime Studies* 13(1), 13.

Gallic, B. L. and A. Cox (2006). 'An economic analysis of illegal, unreported and unregulated (IUU) fishing: Key drivers and possible solutions', *Marine Policy* 30(6), 689–695.

Garaway, C., R. Arthur, B. Chamsingh, P. Homekingkeo, K. Lorenzen, B. Saengvilaikham and K. Sidavong (2006). 'A social science perspective on stock enhancement outcomes: Lessons learned from inland fisheries in southern Lao PDR', *Fisheries Research* 80(1): 37–45.

Garcia, S. M. and D. J. Doulman (2005). 'FAO's Fisheries Programme and the plan of implementation of the World Summit on Sustainable Development', in S. A. Ebbin, A. Hakon Hoel and A. K. Sydnes (eds) *A Sea Change: The Exclusive Economic Zone and Governance Institutions for Living Marine Resources*. Dordrecht: Springer, 169–193.

Grafton, Q. R., R. Arnason, T. Bjørndal, D. Campbell, H. F. Campbell, C. W. Clark, R. Connor, D. P. Dupont, R. Hannesson, R. Hilborn, J. E. Kirkley, T. Kompas, D. E. Lane, G. R. Munro, S. Pascoe, D. Squires, S. I. Steinshamn, B. R. Turris and Q. Weninge (2006). 'Incentive-based approaches to sustainable fisheries', *Canadian Journal of Fisheries and Aquatic Sciences* 63(3), 699–710.

Gulbrandsen, L. H. (2010). *Transnational Environmental Governance: The Emergence and Effects of the Certification of Forest and Fisheries*. Cheltenham: Edward Elgar.

Hakon Hoel, A., A. K. Sydnes and S. A. Ebbin (2005). 'Ocean governance and institutional change', in S. A. Ebbin, A. Hakon Hoel and A. K. Sydnes (eds) *A Sea Change: The Exclusive Economic Zone and Governance Institutions for Living Marine Resources*. Dordrecht: Springer, 3–16.

Hall, D. (2003). 'The international political ecology of industrial shrimp aquaculture and industrial plantation forestry in Southeast Asia', *Journal of Southeast Asian Studies* 34(2), 251–264.

Hampton, J., A. M. Berger, S. Harley, G. M. Pilling and N. Davies (2012). 'Introduction to harvest control rules for WCPO tuna fisheries', MOW1-IP/06. Manila: WCPFC.

Held, D., A. McGrew, D. Goldblatt and J. Perraton (1999). *Global Transformations. Politics, Economics and Culture*. Cambridge: Polity Press.

Hilborn, R. and C. J. Walters (1992). 'Quantitative fisheries stock assessment: Choice, dynamics and uncertainty', *Reviews in Fish Biology and Fisheries* 2(2), 177–178.

Hilborn, R., J. M. Orensanz and A. M. Parma (2005). 'Institutions, incentives and the future of fisheries', *Philosophical Transactions of the Royal Society B: Biological Sciences* 360(1453), 47–57.

HLPE (2014). 'Sustainable fisheries and aquaculture for food security and nutrition', report by the High Level Panel on Food Security and Nutrition of the Committee on World Food Security. Rome: CFS/HLPE.

ICTSD (2011). 'Panel rules US "dolphin safe" label too trade restrictive', *Bridges Weekly Trade News Digest*, 15(31).

ICTSD (2014). 'EU Commission proposes ban on Sri Lankan fish imports, reprieve for Belize', *BioRes News*, 20 October.

ICTSD (2015). 'EU warns Thailand on illegal fishing', *BioRes*, 9(4), 22.

Jänicke, M. (1990). *State Failure: The Impotence of Politics in Industrial Society*. University Park, PA and Cambridge: Penn State University Press/Polity Press.

Jentoft, S. (2000). 'The community: A missing link of fisheries management', *Marine Policy* 24(1), 53–60.

Kastner, J. J. and R. Pawsey (2002). 'Harmonising sanitary measures and resolving trade disputes through the WTO-SPS framework. Part I: A case study of the US–EU hormone-treated beef dispute', *Food Control* 13, 49–55.

Kelleher, K. (2005). *Discards in the World's Marine Fisheries. An Update*. Rome: FAO.

Lack, M. (2007). *Catching On? Trade-Related Measures as a Fisheries Management Tool*. Cambridge: Traffic International.

Le Manach, F., M. Andriamahefazafy, S. Harper, A. Harris, G. Hosch, G.-M. Lange, D. Zeller and U. R. Sumaila (2013). 'Who gets what? Developing a more equitable framework for EU fishing agreements', *Marine Policy* 38, 257–266.

Leroy, A., F. Galletti and C. Chaboud (2016). 'The EU restrictive trade measures against IUU fishing', *Marine Policy* 64, 82–90.

Lipschutz, R. and C. Fogel (2002). '"Regulation for the rest of us?" Global civil society and the privatization of transnational regulation', in R. Hall and T. Biersteker (eds) *The Emergence of Private Authority in Global Governance*. Cambridge: Cambridge University Press, 115–140.

Mace, F. C. (1994). 'The significance and future of functional analysis methodologies', *Journal of Applied Behavior Analysis* 27(2), 385–392.

Maciel, C. T. (2015). 'Public morals in private hands', PhD thesis, Wageningen University, Wageningen.

Mansfield, B. (2007). 'Articulation between neoliberal and state-oriented environmental regulation: Fisheries privatization and endangered species protection', *Environment and Planning A* 39(8), 1926–1942.

Markus, T. (2010). 'Towards sustainable fisheries subsidies: Entering a new round of reform under the Common Fisheries Policy', *Marine Policy* 34(6), 1117–1124.

May, C. K. (2008). 'Achieving sustainability in US fisheries: Community engagement in co-management', *Sustainable Development* 16(6), 390–400.

Mbithi Mwikya, S. (2006). 'Fisheries access agreements: Trade and development issues', Issue Paper No. 2. Geneva: International Centre for Trade and Sustainable Development.

Miller, A. M. and S. R. Bush (2015). 'Authority without credibility? Competition and conflict between ecolabels in tuna fisheries', *Journal of Cleaner Production* 107, 137–145.

Miller, D. D. and U. R. Sumaila (2016). 'IUU fishing and impact on the seafood industry', in A. Naaum and R. Hanner (eds) *Seafood Authenticity and Traceability*. San Diego, CA: Academic Press, 83–95.

Mora, C., R. A. Myers, M. Coll, S. Libralato, T. J. Pitcher, R. U. Sumaila, D. Zeller, R. Watson, K. J. Gaston and B. Worm (2009). 'Management effectiveness of the world's marine fisheries', *PLoS Biol* 7(6), e1000131.

Motaal, D. A. (1999). The agreement on Technical Barriers to Trade, the Committee on Trade and the Environment and ecolabelling', in G. P. Sampson and W. B. Chambers (eds) *Trade, Environment, and the Millennium*. Tokyo: United Nations University Press, 223–238.

Neitzel, S. M., P. A. van Zwieten, A. Hendriksen, D. Duggan and S. R. Bush (2017). 'Returning information back to fishers: Graphical and numerical literacy of small-scale Indonesian tuna fishers', *Fisheries Research* 196, 96–105.

Nielsen, J. R., P. Degnbol, K. K. Viswanathan, M. Ahmed, M. Hara and N. M. R. Abdullah (2004). 'Fisheries co-management – An institutional innovation? Lessons from South East Asia and Southern Africa', *Marine Policy* 28, 151–160.

OECD (2010). *Globalisation in Fisheries and Aquaculture; Opportunities and Challenges*. Paris: OECD.

OECD (2015). 'Green growth in fisheries and aquaculture', OECD Green Growth Studies. Paris: OECD Publishing.

Oosterveer, P. (2006). 'Globalization and sustainable consumption of shrimp: consumers and governance in the global space of flows', *International Journal of Consumer Studies* 30(5), 465–476.

Oosterveer, P. (2008). 'Governing global fish provisioning: Ownership and management of marine resources', *Ocean & Coastal Management* 51, 797–805.

Österblom, H. and Ö. Bodin (2012). 'Global cooperation among diverse organizations to reduce illegal fishing in the Southern Ocean', *Conservation Biology* 26(4), 638–648.

Páez-Osuna, F., A. Gracia, F. Flores-Verdugo, L. P. Lyle-Fritch, R. Alonso-Rodrí, A. Roque and A. C. Ruiz-Fernández (2003). 'Shrimp aquaculture development and the environment in the Gulf of California ecoregion', *Marine Pollution Bulletin* 46(7), 806–815.

Paul, A. (2005). 'Rise, fall, and persistence in Kadakkodi: An enquiry into the evolution of a community institution for fishery management in Kerala, India', *Environment and Development Economics* 10(1), 33–51.

Pinkerton, E. W. (1994). 'Local fisheries co-management: A review of international experiences and their implications for salmon management in British Columbia', *Canadian Journal of Fisheries and Aquatic Sciences* 51(10), 2363–2378.

Pomeroy, R., J. Parks, R. Pollnac, T. Campson, E. Genio, C. Marlessy, E. Holle, M. Pido, A. Nissapa and S. Boromthanarat (2007). 'Fish wars: Conflict and collaboration in fisheries management in Southeast Asia', *Marine Policy* 31(6), 645–656.

Ponte, S., I. Kelling, K. S. Jespersen and F. Kruijssen (2014). 'The blue revolution in Asia: Upgrading and governance in aquaculture value chains', *World Development* 64, 52–64.

Potts, T. and M. Haward (2007). 'International trade, eco-labelling, and sustainable fisheries – Recent issues, concepts and practices', *Environment, Development and Sustainability* 9(1), 91–106.

Pramod, G., K. Nakamura, T. J. Pitcher and L. Delagran (2014). 'Estimates of illegal and unreported fish in seafood imports to the USA', *Marine Policy* 48, 102–113.

Pritchard, B. (2009). 'The long hangover from the second food regime: A world-historical interpretation of the collapse of the WTO Doha Round', *Agriculture and Human Values* 26(4), 297–307.

Prompoj, W. (2013). 'Thailand experience on aquaculture certification', paper presented at workshop on Aquaculture Certification, Ho Chi Minh City, 26 June, available at http://daotao.vasep.com.vn/Uploads/image/Nguyen-Thi-Thanh/file/Thailand%20Experience%20on%20Aquaculture%20Certification-revised.pdf (accessed 21 November 2018).

Reilly, A. and F. Käferstein (1997). 'Food safety hazards and the application of the principles of the hazard analysis and critical control point (HACCP) system for their control in aquaculture production', *Aquaculture Research* 28(10), 735–752.

Samerwong, P., S. R. Bush and P. Oosterveer (2018). 'Implications of multiple national certification schemes for shrimp aquaculture in Thailand', *Aquaculture* 493, 319–327.

Sardà, F., M. Coll, J. J. Heymans and K. I. Stergiou (2015). 'Overlooked impacts and challenges of the new European discard ban', *Fish and Fisheries* 16(1), 175–180.

Schmidt, C.-C. (2004). 'Addressing illegal, unreported and unregulated (IUU) fishing', paper presented at International Fisheries Compliance Conference, Brussels, 29–30 September, available at www.oecd.org/greengrowth/fisheries/34029751.pdf (accessed 16 November 2018).

Schmidt, C.-C. (2017). *Issues and Options for Disciplines on Subsidies to Illegal, Unreported and Unregulated Fishing*. Geneva: International Centre for Trade and Sustainable Development.

Sen, S. and J. Raakjaer Nielsen (1996). 'Fisheries co-management: A comparative analysis', *Marine Policy* 20(5), 405–418.

Shaw, M. (2000). *Theory of the Global State. Globality as an Unfinished Revolution*. Cambridge: Cambridge University Press.

Smith, M. D., C. A. Roheim, L. B. Crowder, B. S. Halpern, M. Turnipseed, J. L. Anderson, F. Asche, L. Bourillón, A. G. Guttormsen and A. Khan (2010). 'Sustainability and global seafood', *Science* 327(5967), 784–786.

Squires, D., M. Maunder, R. Allen, P. Andersen, K. Astorkiza, D. Butterworth, G. Caballero, R. Clarke, H. Ellefsen, P. Guillotreau, J. Hampton, R. Hannesson, E. Havice, M. Helvey, S. Herrick, K. Hoydal, V. Maharaj, R. Metzner, I. Mosqueira, A. Parma, I. Prieto-Bowen, V. Restrepo, S. F. Sidique, S. I. Steinsham, E. Thunberg, I. del Valle and N. Vestergaard (2016). 'Effort rights-based management', *Fish and Fisheries* 18(3), 440–465.

Srinivasan, J. T. (2005). 'State regulation versus co-management: Evidence from the Cochin estuarine fisheries in India', *Environment and Development Economics* 10(1), 97–117.

Sumaila, U. R., C. Bellmann and A. Tipping (2014). 'Fishing for the future: Trends and issues in global fisheries trade', E15Initiative. Geneva: International Centre for Trade and Sustainable Development/World Economic Forum.

Sumaila, U. R., V. Lam, F. Le Manach, W. Swartz and D. Pauly (2013). 'Global fisheries subsidies', Note. Brussels: European Parliament.

Sumaila, U. R., A. Khan, L. Teh, R. Watson, P. Tyedmers and D. Pauly (2010a). 'Subsidies to high seas bottom trawl fleets and the sustainability of deep-sea demersal fish stocks', *Marine Policy* 34(3), 495–497.

Sumaila, U. R., A. S. Khan, A. J. Dyck, R. Watson, G. Munro, P. Tydemers and D. Pauly (2010b). 'A bottom-up re-estimation of global fisheries subsidies', *Journal of Bioeconomics* 12(3), 201–225.

Sydnes, A. K. (2005). 'Regional fisheries organisations and international fisheries governance', in S. A. Ebbin, A. Hakon Hoel and A. K. Sydnes (eds) *A Sea Change: The Exclusive Economic Zone and Governance Institutions for Living Marine Resources*. Dordrecht: Springer, 117–135.

Tran, N., C. Bailey, N. Wilson and M. Phillips (2013). 'Governance of global value chains in response to food safety and certification standards: The case of shrimp from Vietnam', *World Development* 45, 325–336.

UN (2013). 'Sustainable fisheries', resolution adopted by the General Assembly, 9 December, UN Doc. A/RES/68/71. New York: United Nations.

Vogel, D. (2000). 'The environment and international trade', *Journal of Policy History* 12(Special Issue 1), 72–100.

Walters, C. (1997). 'Challenges in adaptive management of riparian and coastal ecosystems', *Conservation Ecology* 1(2).

WCPFC (2014). *Summary Report: Scientific Committee Tenth Regular Session. Commission for the Conservation and Management of Highly Migratory Fish Stocks in the Western and Central Pacific Ocean*. Majuro, Marshall Islands: Western and Central Pacific Fisheries Commission.

Webster, D. G. (2011). 'The irony and the exclusivity of Atlantic bluefin tuna management', *Marine Policy* 35(2), 249–251.

Wilson, D. C., M. Ahmed, S. V. Siar and U. Kanagaratnam (2006). 'Cross-scale linkages and adaptive management: Fisheries co-management in Asia', *Marine Policy* 30(5), 523–533.

WorldBank and FAO (2009). *The Sunken Billions. The Economic Justification for Fisheries Reform*. Washington, DC and Rome: WorldBank/FAO.

Worm, B., E. B. Barbier, N. Beaumont, J. E. Duffy, C. Folke, B. S. Halpern, J. B. Jackson, H. K. Lotze, F. Micheli and S. R. Palumbi (2006). 'Impacts of biodiversity loss on ocean ecosystem services', *Science* 314(5800), 787–790.

Worm, B., R. Hilborn, J. K. Baum, T. A. Branch, J. S. Collie, C. Costello, M. J. Fogarty, E. A. Fulton, J. A. Hutchings, S. Jennings, O. P. Jensen, H. K. Lotze, P. M. Mace, T. R. McClanahan, C. Minto, S. R. Palumbi, A. M. Parma, D. Ricard, A. A. Rosenberg, R. Watson and D. Zeller (2009). 'Rebuilding global fisheries', *Science* 325(5940), 578–585.

WTO (2001). Ministerial Declaration, WT/MIN(01)/DEC/1. Doha: WTO.

WTO (2003). 'European Communities – Trade description of sardines', WT/DS231/18, WTO, available at www.wto.org/english/tratop_e/dispu_e/cases_e/ds231_e.htm (accessed 18 November 2018).

Young, N. and R. Matthews (2011). *The Aquaculture Controversy in Canada: Activism, Policy, and Contested Science*. Vancouver, BC: UBC Press.

5 Governing seafood through markets

Introduction

In light of the challenges faced by governments to establish well-functioning global sustainable seafood governance arrangements, a range of 'market-based', 'non-state' or 'private' initiatives have emerged (largely) in OECD countries. These initiatives vary considerably, but hold in common their ambition to use price signals and market access as incentives for changing practices towards 'better', 'responsible' or 'sustainable' forms of seafood production and trade (Parkes et al., 2010). As we outline in this chapter, the specific ways in which these initiatives have used markets to create change have evolved. Initially relying upon boycotts, over time the NGOs behind these initiatives have created a sophisticated portfolio of ecolabels, recommendation lists and improvement projects across an ever-expanding spectrum of market claims – these now include fair, local and organic (Roheim, 2009; Jacquet et al., 2010).

Behind the expansion of these tools and arrangements has been a series of partnerships between NGOs and (predominantly) US-based family philanthropic foundations. Indeed, the global nature of seafood trade, but more narrowly defined as exports to the US and EU, has been perceived as an opportunity for affecting ocean conservation in nations that might not otherwise be influenced to improve environmental practices. Common to all these tools is the goal of communicating 'sustainability' as a credence product attribute – that is, one that is not discernible through experience but instead through verification by a trusted third party (Darby and Karni, 1973). These tools build on the broader phenomenon of political or ethical consumerism (Stolle and Micheletti, 2013), which aims to alter demand for products with and without sustainability attributes and thereby create economic incentives for producers to improve their production practices. However, as we demonstrate, the motivation of NGOs to develop these initiatives is to embed themselves further in state-led decision-making around both national and international rules and regulations.

This chapter analyses this rapidly expanding domain of market-based voluntary seafood governance initiatives introduced alongside and beyond national governmental regulations. We first give an overview of these initiatives and then discuss how they relate to governmental regulations, to fishers included or excluded from the schemes, to the seafood industry, and to the retailers and consumers. We conclude the chapter with a discussion on the contributions these voluntary market-based approaches can offer to sustainable seafood provision.

Creating market-based change

Demand driving supply

Over the last 30 years civil society organisations have initiated several sustainable seafood governance arrangements. Anywhere between 30 and 50 voluntary labelling and certification schemes have been introduced by NGOs and private companies, while other NGOs introduced consumer guides or started boycott campaigns (Parkes et al., 2010; Potts et al., 2016). These arrangements are generally market based, i.e. they try to create incentives within the existing seafood supply chain to promote more sustainable fisheries' and aquaculture practices by shaping market demand (Gudmundsson and Wessells, 2000). Their stated intention is to change seafood production and trade practices by encouraging consumers to take sustainability more into consideration when buying seafood and thereby to make it attractive for retailers to supply sustainably sourced produce. As outlined by Sutton (1998), growing market demand for more sustainable seafood was thought to contribute to expanding sustainability considerations in fisheries and aquaculture and 'trickle down' (or across) to the sector as a whole, including improved government regulation.

These arrangements are premised on the observation that global seafood supply is based on 'a series of interconnected flows of goods, services, incentives, and information between the different participants involved' (Gagalyuk et al., 2010: 140). The leverage of these interconnected material and immaterial flows involved in global seafood supply networks can be used either by those working in the industry to produce and trade fish or for the promotion of sustainable seafood. These chains and networks become, in other words, sites of sustainability governance (Bush et al., 2015). This means that increased global trade not only enhances the complexity and volume of seafood supply networks but also opens up opportunities for societal actors to leverage change towards improved production, trade and consumption practices. As chains become more integrated with digital technologies, the possibilities for steering global seafood supply networks towards more sustainability continue to grow.

Comparable to other food products, there has been an apparent shift in the seafood sector from a supplier-driven to a buyer-driven orientation in the structuration of the supply chains (Wilkinson, 2006). This shift signifies the strengthened roles of the retailing sector in organising the supply chain, including the choices of consumers. In OECD countries this position is demonstrated by the fact that supermarkets sell up to 90 per cent of all available seafood, the majority sourced from developing countries (Kelling, 2008). For example, Walmart (the world's largest retail chain) sources seafood from 70 countries, while the UK supermarket chain Tesco sources from 38 different ones (ibid.). Beyond the OECD this is also a growing trend. Many US and EU retailers are multinationals, meaning they have penetrated many Asian, African and Latin American countries.

The rise of these retailers is of central importance to the rise of so-called market-based forms of seafood governance because of their buyer-driven control over value chains. The pledges made by retailers (and other corporations involved in seafood supply networks) in major OECD markets to adopt voluntary sustainability certification schemes have driven the use of these approaches. The reasons for adoption vary, but come down to the need to secure their future seafood supply, promote sustainable production and protect their reputation in the eyes of their customers from being damaged by adverse publicity. It is the overall vulnerability of supply and market positions on which NGO-led market-based interventions have focused. In short, the theory of change behind these approaches can be labelled 'demand driving supply': NGOs shape the demands retailers place upon their suppliers to deliver credence-related product attributes like sustainability.

From boycotts to buycotts

The first attempts by NGOs to change market demand towards sustainable forms of seafood focused on consumer boycotts of high-profile fish species in the US. Perhaps the most prominent of these NGO campaigns was the consumer backlash against tuna purse-seine fisheries in the Eastern Pacific Ocean in which dolphins were entangled in tuna nets, leading to maiming, drowning and death. Following widespread media coverage in the US, schoolchildren boycotted tuna in their school lunches, and further public pressure placed upon the three major tuna companies, Chicken of the Sea, Starkist and Bumble Bee, led to a boycott on purchasing tuna caught using means that were harmful to dolphins (Wessells and Wallström, 1994).

Other prominent examples of boycotts in the 1990s included the 'Give Swordfish a Break' and 'Take a Pass on Chilean Sea Bass' (otherwise known as Patagonian toothfish) campaigns. Both boycotts included public media campaigns targeting consumers through sophisticated communications strategies, and enlisting the consumer to pressure chefs, restaurants, retailers and even cruise ships to cease buying swordfish and Chilean sea bass (see Roheim and Sutinen, 2006a; Jacquet and Pauly, 2007). The Chilean sea bass campaign successfully led to retailers and restaurants pulling the product from their shelves and menus. It helped focus attention on the thread of IUU fishing and trade (Roheim and Sutinen, 2006a), and also led to the improved status of Chilean sea bass stocks. In contrast, the swordfish campaign did not lead to reduced imports to the US, but did incite the US government to close several areas for fishing swordfish, institute stronger quota restrictions, improve monitoring of imports of swordfish for undersized fish and create improved management measures to reduce capture of undersize fish (Roheim and Sutinen, 2006b).

Boycotts of aquaculture products have focused mainly on shrimp, one of the most important as well as most controversial seafood products on the global market (Islam, 2014). In 2009 the Mangrove Action Project launched the 'Question Your Shrimp' campaign, leading to the collection of 1,000 pledges from people agreeing not to eat imported shrimp and 33 restaurants and two retailers in Seattle which agreed not to sell imported shrimp (MAP, 2017). The aim of the campaign was to reduce sales in the US, which is the largest importer of shrimp, and by reducing demand to stop the deforestation of coastal mangroves. The campaign worked to influence consumers, but also enrolled retail outlets by having them joined discursively in the problem of mangrove deforestation, in addition to the threat of further consumer action.

Perhaps the most successful seafood boycott was the Greenpeace 'Recipe for Disaster' campaign in the UK in 2004. This mediagenic campaign placed direct pressure on highly competitive retailers to stop sourcing species which were on the International Union for the Conservation of Nature red list, species with high levels of bycatch and vulnerable skates and rays (Greenpeace, 2005). The boycott involved, among other direct actions, Greenpeace signage and dumping bycatch from beam trawlers on the steps of retailers as part of what they labelled 'the rancid roadshow'. Within a very short period the UK's top ten retailers agreed to make changes to their seafood sourcing policies. Soon after this, similar campaigns were coordinated by Greenpeace and the WWF in the rest of the EU and the US. However, unlike the other boycott campaigns, the outcome was a series of retail commitments to source only seafood certified by the MSC and ASC or recommended by seafood lists such as Monterey Bay Aquarium's Seafood Watch. The role of *boycotts*, therefore, appears to have been usurped by varying forms of *buycotts* – campaigns to encourage consumers to buy particular products deliberately for ethical, social or environmental reasons (Stolle and Micheletti, 2013).

The first recommendation lists started with the 1998 *Audubon* list of ranked seafood, intended to provide consumers with information to seek out 'good choices' and avoid 'bad choices' in their seafood purchasing decisions (Safina, 1998). With funding from major private foundations in various parts of the world, these lists have proliferated. Today the best-known guide is perhaps that of Monterey Bay Aquarium under the Seafood Watch programme. Aquariums like Monterey Bay have expanded their educational remit beyond their initial focus on life in the oceans to the broader perspective of ocean health. For many of these aquariums, the health of fisheries as the source of seafood was seen as a natural extension (Kemmerly and Macfarlane, 2009). One of the formats that this extended seafood education programme took was pocket guides or wallet cards (and now smartphone apps), aimed at reaching and educating consumers. Seafood guides provide a rating using a criteria-based methodology that evaluates environmental and biological criteria of species, fisheries or aquaculture practices. The rankings are summarised in a traffic-light system of red (items to avoid), yellow (good alternatives) and green (best choices), which can be used to guide consumers to purchase fish products that are well managed or farmed.

Of the 200 or so guides produced internationally, many were initially independently researched, resulting in conflicting recommendations and consumer confusion (Roheim, 2009). Many of these recommendation lists have also come under criticism for not providing empirical assessments of fish stocks and aquaculture systems, and not providing clear evidence that they have led to positive impacts on the sustainability of these systems (Jacquet and Pauly, 2007; Kemmerly and Macfarlane, 2008) or to changes in consumer demand (Hallerstein and Villas-Boas, 2013). Others have questioned whether the impacts of these recommendation lists are related more to the opportunities they afford the NGOs and aquariums running the schemes to gain access and authority in formal policy discussions (de Vos and Bush, 2011). Nevertheless, these lists still play an important role in the US, the UK and the Netherlands, given their perceived independence and therefore the credibility of the organisations that are promoting them.

Labelling and certification schemes

Seafood ecocertification programmes are perhaps the most complex, costly and controversial form of market-based assessment and verification of fisheries' and aquaculture performance. While there is variation between programmes, they generally involve (often multiple) auditors empirically assessing the performance of producers, regulatory bodies and value-chain actors against an independent standard. If they pass these assessments, certified producers are permitted to place ecolabels on affiliated products, serving as a signal to consumers that the product was produced from a well-managed source (Ward and Phillips, 2008). In return, consumers purchasing labelled seafood create a direct incentive, such as a price premium, for sustainable fishing (Gudmundsson and Wessells, 2000). How these certification schemes emerged and what types of engagement they engender with the seafood industry are best presented by dividing out fisheries and aquaculture.

Fisheries' certification

The number of these certifications has increased since 2000 to include a range of sustainable, responsible, organic and 'fair' standards and labels. The volume and scope of fisheries that are certified have also increased over time (see Figure 5.1 and Box 5.1). In 2015 it was estimated that three certification schemes – the MSC, Friends of the Sea (FoS) and Naturland Organic – had certified around 20 per cent of total global landings (Potts et al., 2016). While impressive given their rapid growth, these volumes alone do not say everything

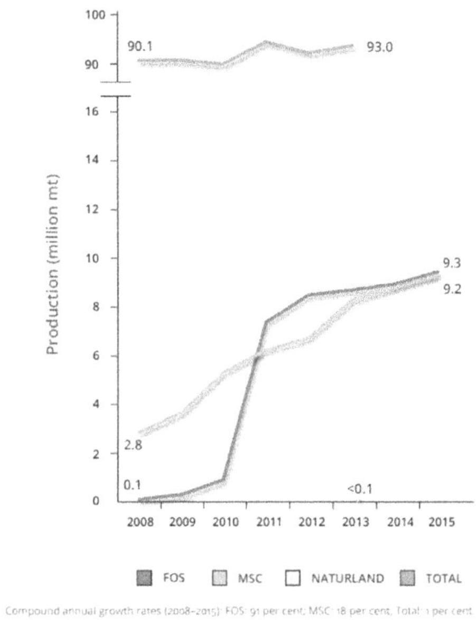

Figure 5.1 Increases in volumes of certified fisheries.

Source: Potts (2016), used with permission.

about the impact that these certification schemes are having. Based on this impact, as well as how their standards are defined and assessed, the certification schemes vary in their level of perceived credibility and authority in the market. The following paragraphs outline a short history of fisheries' certification, with a focus on the most prominent schemes.

Box 5.1 FAO-developed guidelines for ecolabelling of fish and fishery products from marine-capture fisheries

To reduce the confusion resulting from the emergence of multiple labels, the FAO developed guidelines for the ecolabelling of fish and fishery products from marine-capture fisheries. These guidelines are based on the following principles.

- Be consistent with relevant international agreements, including UNCLOS 1982, the FAO Code of Conduct for Responsible Fisheries and the WTO rules.
- Recognise the sovereign rights of states and comply with all relevant laws and regulations.
- Be of a voluntary nature and market driven.
- Be transparent.

- Be non-discriminatory, do not create obstacles to trade and allow for fair competition.
- Provide the opportunity to enter international markets.
- Establish clear accountability for the owners of schemes and the certifying bodies in conformity with international standards.
- Incorporate reliable auditing and verification procedures.
- Be considered equivalent if consistent with these guidelines.
- Be based on the best scientific evidence available, also taking into account traditional knowledge of resources provided that its validity can be objectively verified.
- Be practical, viable and verifiable.
- Ensure that labels communicate truthful information.
- Provide for clarity.

Source: FAO (2009)

The EII Dolphin Safe label came about in the 1980s after media attention was given to the practice of setting purse-seine nets on dolphins, known to associate with tuna in the tropical Eastern Pacific Ocean. In the 1970s and 1980s hundreds of thousands of dolphin deaths were associated with this fishing practice (Hall and Boyer, 1986; Baird and Quastel, 2011). In response, the EII, an environmental NGO, launched a campaign aimed at creating consumer awareness of the plight of the dolphins and its Dolphin Safe label. Although targeted at the US market initially, major retailers in other parts of the world also started to display logos ensuring their tuna was 'dolphin safe' or 'dolphin friendly' (Brown, 2005). This meant the certification had gone beyond the canned tuna product to encompass the entire supply chain, giving the EII a high degree of network power. To date over 500 companies are certified dolphin safe, accounting for 90 per cent of the market and covering 71 countries.[1]

For tuna to be considered 'dolphin safe' the company selling the tuna must declare that it undertakes no intentional chasing, netting or encirclement of dolphins during an entire tuna fishing trip; no use of drift gill nets to catch tuna; no accidental killing of or serious injury to any dolphins during net sets; and no mixing of dolphin-safe and dolphin-deadly tuna in individual boat wells (for accidental killing of dolphins), or in processing or storage facilities. Furthermore, each trip in the tropical Eastern Pacific Ocean by vessels of 400 gross tons and above must have an independent observer on board to attest to compliance with the standards (EII, 2012). These criteria have been updated, and include a ban on IUU fishing vessels and the statement that companies 'should not engage in shark finning' (EII, 2011: 5). While providing a classification of what constitutes 'dolphin safe', no procedural information on the certification process itself is given.

The 'Dolphin Safe tuna' label was the first successful seafood labelling scheme. However, it is a single-attribute label, focusing on the protection of one single mammal species, dolphins, in one single fishery, tuna (Thrane et al., 2009). While the dolphin issue in the 1980s and 1990s provided a credible basis upon which to build the label, the current relevance and credibility have been questioned. Miller and Bush (2015) argue that the label lacks a coherent and consistent system of standards and criteria for the assessment procedure, reverts monitoring to 'self-reporting skippers', has poor transparency and offers little incentive for improving practices within the fishery as a whole. They therefore conclude that complying with the Dolphin Safe standard represents the lowest common denominator of sustainability, given it does not require a company to make any improvements to its practices to achieve certification. Certification schemes developed since Dolphin Safe have generally gone beyond single issues to address issues such as overfishing and the overall protection of marine species and habitats.

Arguably the most advanced standard for capture fisheries, the MSC, established in 1997, has the longest and broadest presence in the market relative to other ecolabels (Parkes et al., 2010; Stemle et al., 2016). The MSC fisheries standard has three overarching principles. Principle 1 sets out the requirement for a fishery to be 'conducted in a manner that does not lead to over-fishing or depletion of the exploited populations and, for those populations that are depleted, the fishery must be conducted in a manner that demonstrably leads to their recovery'. Principle 2 focuses on the environmental impact of fishing – ensuring that fishing operations allow 'for the maintenance of the structure, productivity, function and diversity of the ecosystem (including habitat and associated dependent and ecologically related species) on which the fishery depends'. Principle 3 focuses on the effectiveness of the management regime in place around a fishery, prescribing that the management system 'respects local, national and international laws and standards and incorporates institutional and operational frameworks that require the use of the resource to be responsible and sustainable' (MSC, 2014).

The scientific credibility of the MSC involves these three principles plus more specific criteria and performance indicators used to assess the fisheries independently. Each of the three principles is broken down into 31 performance indicators, which represent the sustainability of a fishery under assessment and are therefore the fundamental determinants of credibility. Performance indicators are based on three 'scoring guideposts': an 'ideal' fishery would score 100; a 'best practice' fishery would score 80; and the conditional level of entry into the MSC certification procedure is 60. To become certified the weighted average of all performance indicators must achieve a score of 80 or more for each of the principles. The MSC principles are translated into a detailed management plan for a specific fishery to assure its responsible and sustainable use of the fish stock, and the certification process is carried out by an independent auditor.

By the end of 2017 a total of 312 fisheries had received the MSC label, most of which are in OECD countries except for some certified fisheries in South Africa, Vietnam and Mexico, while another 68 fisheries were undergoing assessment.[2] Fisheries in over 30 countries were engaged with MSC certification, involving over 9.5 million MT of capture fisheries' production, representing 12 per cent of wild-caught seafood landings globally. In supermarkets and specialised fish retail stores in 66 (mostly industrialised) countries, currently more than 25,000 different MSC-labelled items from certified fisheries are available, while US$4.8 billion worth of products were labelled. With the MSC label, a new 'green' product identity is created in an otherwise anonymous global fish market.

The MSC initiative was followed some years later by other initiatives introducing sustainable seafood labels, among which the FoS label is the most prominent. FoS was founded in 2006 as an international sustainability certification project for products originating from fisheries and aquaculture. The FoS Approved Fisheries standards certify target stocks which are not overexploited and use of fishing methods which do not impact the seabed and generate less than 8 per cent discards on average. By August 2017 111 fisheries and 100 aquaculture farms were certified, with 323 FoS-certified products available in 70 countries.[3] According to Potts et al. (2016), FoS certifies 8.6 million MT of global capture fisheries' production, corresponding to approximately 10 per cent of total seafood production (based on the FAO's 2013 data).

The number of certification schemes for fisheries goes beyond these major ecolabels, but many of the additional schemes have a marginal market share and are focused on specific issues. They include the Responsible Vessel Scheme, launched in 2016 to certify fishing companies' performance concerning 'health, safety and crew welfare, training and professional development, the vessel and its mission, care of the catch, and care for the environment'.[4] As we outline in the next chapter, other forms of certification have emerged for fisheries that go beyond the claims of sustainability – both in competition with and as a complement to these labels.

Aquaculture certification

Also aquaculture certification has increased dramatically over the last two decades. The major private third-party certification schemes are the Best Aquaculture Practices (BAP) standards of the GAA and the ASC. A wide range of other private labels has also emerged, including the business-to-business GlobalGAP scheme and organic labels like Naturland. But unlike the fisheries' sector, there is a range of national good aquaculture practice and better management practice standards developed by national governments, particularly from Southeast Asia. The volumes of certified aquaculture, estimated at 3 per cent of total global seafood production, are overall less that than that of capture fisheries, which represent 11 per cent of global seafood production (based on Potts et al., 2016 – see Figures 5.1 and 5.2 combined). However, like capture fisheries, it remains unclear how much this coverage will grow. For example, Bush et al. (2013a) argue that the further expansion

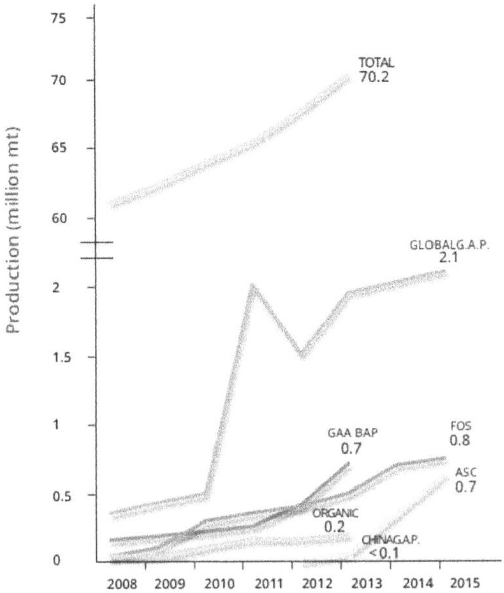

Compound annual growth rates (2008–2015): ASC: 98 per cent; BAP: 35 per cent; FoS: 47 per cent; GLOBALGAP: 29 per cent; Organic: 35 per cent; Total: 6 per cent. Sources: FAO Fishstat, 2015; ASC, BAP, ChinaGAP, FiBL, FoS, GLOBALGAP, MSC, Naturland, personal communication, 2015.

Figure 5.2 Increases in volumes of certified aquaculture.
Source: Potts (2016), used with permission.

of third-party aquaculture certification may be limited by market demand, which they estimate to be limited to the equivalent of 14 per cent of global production, and the weak capability of the majority of producers in developing countries to comply with standards.

Certifying aquaculture differs from certifying fisheries, because where fisheries' labelling and certification schemes mainly address ecosystems' and fish stocks' management, aquaculture schemes deal with the use of fishmeal, the application of antibiotics (resulting in concerns about food safety and water pollution), the impacts on biodiversity (destroying mangroves when putting fishponds in coastal zones) and labour conditions in the processing industry. As aquaculture covers a wide range of different species as well as different technologies, there is much greater diversity among and within the labelling and certification schemes than is the case for capture fisheries (Boyd et al., 2005). Here we limit our analysis to the most dominant sustainability labels developed for aquaculture-based seafood: the GAA BAP, GlobalGAP and the ASC.

The GAA is an international non-profit trade association founded in 1997 by aquaculture and food companies, mostly from the US. The GAA has developed its industry-based BAP standard as a voluntary certification programme to secure environmental and social responsibility, animal welfare, food safety and traceability in aquaculture. The organisation has introduced multi-species farm standards that apply to all types of production systems for finfish and crustaceans, excluding cage-raised salmonids, which have a separate standard. Unlike other aquaculture certification schemes, BAP has a star rating system for consumer products, representing the certification of different facilities along the value chain. One star represents farm certification only, moving to four stars representing the certification of hatcheries, feed mills and seafood processing and repacking plants.[5] BAP is a standard that allows the final product to be labelled, and the blue BAP-certified label is currently the dominant label in farmed seafood sold in North America.[6] Boyd and McNevin (2012) report that BAP had certified 304,798 MT of channel catfish, pangasius, shrimp and tilapia.

While GAA BAP dominates the US market, the European market has historically been spread between GlobalGAP and the ASC. GlobalGAP was established in 1997 by the Euro-Retailer Produce Working Group (EUREP) to combine auditing food safety standards with specific protocols for sustainability in food production.[7] The Integrated Aquaculture Assurance standard entails meeting criteria for legal compliance, food safety, animal and workers' welfare, and environmental and ecological care. The standard includes general requirements that are relevant for all aquaculture farms, supplemented by species-specific standards (salmon and trout, shrimp, pangasius, tilapia and other). By 2007 about one-fifth of all global salmon production was GlobalGAP certified. Boyd and McNevin (ibid.) report that GlobalGAP had certified 500,000 MT of pangasius, salmon, shrimp and tilapia. Most of GlobalGAP-labelled seafood is sold through European retail chains. By 2014 GlobalGAP claimed to have certified a total of two million MT of aquaculture products from 300 companies originating in 30 countries.[8] GlobalGAP is the leading certification scheme for aquaculture by far (Potts et al., 2016).

Twenty years after being initiated as a business-to-business label GlobalGAP adopted a consumer-facing label, first with FoS in 2014 and later with its own GlobalGAP number (GGN) label in 2016. The GGN label is based on the GlobalGAP standard for aquaculture, but in addition provides a consumer-facing coding system on products to make them traceable back to where the fish was produced. The system is also an open platform to other standards, linking a traceable GlobalGAP number to certified products. The only label that has partnered with GGN to date is the FoS, with the aim of increasing the scope of aquaculture products available. FoS, viewed as a more accessible (and less rigorous) standard than BAP or ASC, also allows a broader scope of small-scale aquaculture producers to access GlobalGAP markets in line with the wider smallholder programmes of GlobalGAP.[9]

Building on the positive impacts from MSC in the case of marine fisheries, the WWF took the initiative to develop a comparable scheme for aquaculture, the ASC. The ASC was founded in 2009 by the WWF and IDH

(Netherlands Sustainable Trade Initiative) to introduce global standards for responsible aquaculture based on the conclusions and recommendations from the Aquaculture Dialogues that had been initiated by the WWF some years before. The ASC has developed standards for eight aquaculture species (abalone, bivalve, freshwater trout, pangasius, salmon, shrimp, tilapia and seriola/cobia).[10] Certification according to these standards is conducted by accredited independent third-party certifiers. By September 2017 in total 516 farms were certified while 144 farms were in the process of being certified.[11] These certified farms produced 2,501 certified seafood products available in 44 (mostly Western European) countries.

These schemes differ in many respects regarding content, approach and market, but overall their relevance is increasing. It is challenging to determine their exact scale and impact because of the differences between them and the absence of reliable (independent) and comparable data. Nevertheless, it appears that certified seafood production increased from 0.5 million MT in 2003 to 23 million MT in 2015, representing 14 per cent of global seafood production with a retail value of US$11.5 billion (Potts et al., 2016). These figures show that voluntary schemes have become an established strategy within the sustainable seafood movement, with substantial market shares in the US, Europe and Japan but less in other parts of the world. Nevertheless, the introduction and proliferation of these labels (currently more than 50; ibid.) raise some contentious issues, including their relationships with governments, fishers, industry, retailers and consumers. These relationships are discussed in the following section.

Governments and private voluntary initiatives

Although NGO-led seafood certification schemes are private, voluntary and market based, they can nevertheless not be separated completely from governments. The relationships between them can be characterised as a combination of competition and cooperation.

Private initiatives are competing with governments because they represent an alternative seafood governance arrangement dealing with issues in the global seafood production network that escape the control of national governments. So while governments develop national environmental policies, agree on international environmental agreements and have set up specific regulations for international trade through the WTO (see Chapter 4), private labels and certification schemes are intervening in these fields as well, albeit in different ways.

Competition between governments and private initiatives is particularly relevant for non-OECD countries exporting seafood to OECD countries, as private schemes interfere more directly with national policies, raise issues about the definition of sustainable seafood provision and challenge national sovereignty in political decision-making and the division of labour between public and private authorities. Some governments, especially in Asia, publicly protest against the interference from (mostly Western) NGOs in their domestic affairs (Mbengue, 2011), because they consider this a breach of their national sovereignty. Notably, the lack of legitimacy of civil society organisations when intervening in matters of international trade competes with the officially established role of national governments (Vandergeest and Unno, 2012); for a discussion on the morality of Western labelling for steering farmer practices see Hatanaka (2009). In response, some countries, such as Thailand (Oosterveer, 2006) and Japan (Moye, 2010), have developed their own official national sustainable seafood standards for the global market.

Through the development of national certification schemes, governments aim to attain multiple goals. First, national certification allows the national product to acquire a distinctive position in the market; second, particular claims can be made with respect to the quality, health and sustainability of these products; and third, it can serve as a driver for producers to improve their performance. There are some indications that national labels may be useful on the domestic market but not necessarily on the international one (for an example on poultry labels see de Krom and Mol, 2010). When competing with private initiatives, government labels often lose out in obtaining trust among the general public and in the market. Many consumers in OECD countries consider private schemes to be more credible than most government claims, because the latter are seen as being biased by national interests and less subject to scientific rigour, inclusiveness, transparency and independence compared with private initiatives (Boström, 2006; Auld and Gulbrandsen, 2010).

Government certification not only competes with private initiatives, but may also collaborate with them. For instance, the Dutch government supports the IDH in its activities to promote sustainable trade (Driessen et al., 2012). Backing the introduction of the ASC as a global aquaculture label was one of the IDH's most prominent achievements in recent years. This way, public funds are used to develop and promote a private sustainability certification scheme. In other cases governments consider private voluntary seafood labels an adequate strategy to complement their own limited set of governance tools and collaborate directly with them. Examples are the Western Australian government's announcement of giving financial support to local fisheries to gain MSC certification,[12] the Vietnamese government's conclusion that to be successful in the promotion of pangasius on the global seafood market private regulation was required, and the Canadian government's collaboration with MSC to certify a regional shrimp fishery (Foley, 2013). In 2009 and 2010 the Dutch government subsidised organised fishers to certify their fishery or aquaculture activity as environmentally friendly according to a private scheme.[13]

Through international development aid, governments from OECD countries also support small-scale and developing countries' fisheries that face trouble when trying to get included in private schemes due to the costs and resources involved. Another important but often taken for granted form of cooperation between governments and private schemes is the provision by public authorities of a basic legal infrastructure – clearly defined property rights, the existence of contract law, judicial conflict resolution and freedom of organisation and speech (van Waarden, 2012). Without such legal infrastructure, private initiatives cannot function effectively, which may also (partly) explain why private initiatives are more successful in OECD countries. Governments may also intervene when labels make unsubstantiated claims (Kelling, 2008) to ensure there is 'truth in advertising' and avoid greenwashing by private companies.

Finally, private labels and certification schemes may collaborate with governments as part of a division of tasks. Private initiatives may put sustainability problems on the political agenda or attract public attention by introducing a certification scheme in a seafood GVC, which is then subsequently addressed through official regulation. This way private standards are not replacing governments but 'use their first-mover status to become an additional instrument for rule development and implementation' (ibid.). To a certain extent, the 'dolphin–tuna' case illustrates such a dynamic, as well as shrimp exports from Thailand and Vietnam in some respects.

Producers and labels

The relationship between fishers and private voluntary labels and certification schemes is complex. The schemes offer fishers the possibility to access attractive markets, but they face the problem of lack of involvement in developing the schemes, the costs and complexities of becoming certified, potentially excluding particular categories of fishers, and the limited rewards of being certified.

NGOs and private companies are driving most certification initiatives, but fishers were hardly included in their design. This leads to questions over the legitimacy of these private initiatives (Oosterveer, 2015) and the relevance of the sustainability issues and targeted species included, as standards were developed with concerns from consumers in Western Europe and the US in mind which do not necessarily fit with the fishing practices and sustainability concerns prevailing in producing countries (Standing, 2009; Hatanaka, 2010; Bush et al., 2013b). For instance, ecological issues, especially biodiversity, are considered more important than social issues related to production and trade (Islam, 2014). In addition, certification schemes risk oversimplifying sustainability issues (Belton et al., 2009) and may thereby exclude promising alternative practices.

Smallholders and poor fishing communities often lack the necessary technical capacities, access to historical harvest data and financial and human resources to apply successfully for certification (Gulbrandsen, 2009; Hatanaka, 2009; Belton et al., 2011; Wilkings, 2012). Some question whether in their attempt to cover all relevant issues these standards are not too complicated and too comprehensive for most smallholder fishers (UNEP, 2009). For instance, developing world fisheries currently only make up 7 per cent of the MSC certified fisheries and 9 per cent of the fisheries in full assessment. More than 40 developing world fisheries are, however, under pre-assessment for MSC certification (MSC, 2011). As Cambridge et al. (2011) note, 55 per cent of the fisheries moving through to full assessment still derive from the North Atlantic. One explanation for this bias is the substantial financial investment involved in certification to cover the direct and indirect costs. Direct costs relate to the auditing and application processes, and indirect costs to changing techniques and practices when implementing and maintaining the standard's requirements. These costs may be considerable for smallholders, and this becomes even more relevant when (as is often the case) certification does not translate into price premiums (Wakamatsu and Wakamatsu, 2017). But without certification these smallholders and poor fishers would have no access to attractive consumer markets in Europe and the US at all, and they have therefore little alternative but to accept the additional costs (Bailey et al., 2016). Overall, the marginalisation of small-scale producers in attractive seafood export markets may be unintended but is nevertheless a consequence and limitation of ecolabelling.

Three other important issues are included in the debate on fishers and certification schemes.

First, the effectiveness of certification is questioned because of the inbuilt selection of fisheries and farms and the challenge of achieving continuous improvement. To include small-scale fisheries in developing countries, a pre-assessment phase in certification has now been taken up by a wide range of NGOs under the banner of fishery improvement projects (FIPs) (Deighan and Jenkins, 2015; Sampson et al., 2015). In FIPs, NGOs, private companies and fishery consultancy firms support small-scale (developing world) fisheries to move towards more sustainability in their fishing practices in a stepwise manner.

Second, the introduction of multiple, competing sustainable seafood labels in recent years raises questions about their relative quality and effectiveness, and how fishers should decide on which scheme to select to certify their fishery. It is not necessarily the 'best' label that gets selected, because the requirements for certification, the costs involved and the recognition in particular consumer markets are key criteria for fishermen when they select a label, plus other supply-chain actors (Miller and Bush, 2015; Kirby et al., 2014).

Third, a debate is emerging on the limitations of certifying individual aquaculture farms, because certification is awarded on the basis of their performance but the cumulative sustainability impact from multiple certified farms in the same area might nevertheless be negative (Boyd and McNevin, 2012; Bottema et al., 2018). The same counts for the certification of one fishery as only one element of a complex ecosystem for which more comprehensive sustainable governance regulation is needed. This would require a place-based governance arrangement rather than an individual firm-based certification scheme. However, determining the most appropriate framework for a place-based governance arrangement remains a challenge.

The seafood industry, retailers and labels

Retailers have become the dominant market actors in steering most food supply chains, so they also influence the development and use of labels. Some observers argue that despite claims by certification initiatives that they intend to address consumer concerns, retailers' decisions are more important in determining their success (Auld and Gulbrandsen, 2010). Without support from supermarkets a certification scheme cannot become successful, because retailers decide on which certification scheme to accept in their shops and how to communicate this to the consumer. In some instances large retailers even have a direct influence on the content and organisation of a certification scheme. For example, GlobalGAP is directly controlled by a group of European supermarket chains.

Private certification schemes are criticised from three interrelated views that appear to undermine the market-based assumptions on which they are built. First, it is increasingly unclear if the ecolabel is driving consumers' demand for seafood products in major developed countries like the US and other markets (Bronnmann and Asche, 2016). Second, while a price signal from MSC-labelled products has been observed in several OECD markets (e.g. Roheim et al. 2011; Sogn-Grundvag et al., 2013, 2014; Bronnmann and Asche 2016), there is mixed evidence on whether the premium is being passed through the value chain to producers, who often bear the burden of the costs of certification, thus perhaps failing to incentivise them to bear the costs of changing their production practices (e.g. Blomquist et al., 2015; Stemle et al., 2016). Third, as already outlined, only 7 per cent of MSC-certified fisheries are from developing countries, leaving those regions of the world in need of the largest improvements, and with some of the highest market demand for seafood, out of the scope of the programme (Gulbrandsen, 2009; Ponte, 2012; Bush et al., 2013a).

Retailers may actively promote sustainability in seafood provisioning. For instance, the Dutch national supermarket association CBL introduced in 2008 a plan to create more sustainability in fresh and frozen fish for sale in supermarkets ('*Vis Beter*' in Dutch, or 'Fish Better').[14] In this plan, the Dutch supermarkets agreed that by 2011 all captured seafood sold should comply with the MSC standards or equivalent, but in 2011 this date was postponed to 2016 and the ASC was explicitly included as an accepted scheme for seafood from aquaculture. In 2016 the goal was still not realised. CBL claimed that its original goal was not achieved because there was insufficient supply of certified seafood,[15] but it nevertheless suggested that most captured seafood sold in Dutch supermarkets was MSC certified or comparable.[16] Other retailers and food service companies in Europe and the US have made similar promises (for more information and an overview see Fiorillo, 2014). The problem of how to deal with the presence of multiple sustainability labels was taken up by the Sustainable Seafood Coalition (SSC). The SSC, founded in 2011 by significant seafood suppliers, brands, supermarkets and restaurants in the UK, intends to benchmark seafood labels and address the problem of misleading sustainability claims made by seafood brands on their products. To advance this goal, the SSC developed voluntary codes of conduct on Environmentally Responsible Fish and Seafood Sourcing and on Environmental Claims.[17] A comparable, more recent initiative is the Global Sustainable Seafood Initiative (GSSI), through which seafood companies, mainly from Europe and North America, formed a partnership to increase transparency in international seafood markets.[18] The GSSI developed a benchmarking tool to assess existing certification schemes against a set of baseline requirements. Schemes that meet these are allowed to use the GSSI logo (Ababouch et al., 2016). Recently, the FAO also joined the GSSI.

Labelling and certification schemes are oriented towards global seafood governance. However, national contexts still matter in their development and implementation (Manning et al., 2012). For instance, the presence of multiple seafood labels forces supply-chain actors to choose which one(s) to use in their sourcing and supplying policy. Importers are key switchers in the seafood supply chains because they connect retailers with producers from different parts of the world. It seems importers essentially respond to the demands from large retailers when choosing a particular sustainable seafood label. In a recent study comparing Dutch shrimp importers, Vasilev (2014) found that importers rely on requests from retailers in their decisions on importing certified seafood and which scheme to select. In the Netherlands, retailers' choice of a particular scheme has more to do with avoiding bad publicity than with the actual sustainability claims of the selected scheme. Retailers therefore compare the labels primarily by the reputation of the NGO and other stakeholders backing the certification scheme rather than by their content. This strategy may explain why some labels are more popular in some markets than in others, independent of the scheme's quality or (environmental) effectiveness. Interestingly, retailers were less concerned about logos, as they were afraid of confusion among consumers and ultimately aimed at doing away with logos altogether and relying solely on their company image to convince consumers about the quality, safety and sustainability of the seafood on offer (ibid.).

While the relationship between retailers and importers seems to be without too many complications with regard to sustainability certification, there are some indications of a more conflictual relationship between retailers and seafood suppliers. Critical problems are the distribution of costs involved in fisheries' certification and the division of responsibilities in determining the criteria for sustainable seafood. A clear illustration of these issues is the struggle between the large US retail company Walmart and Alaskan salmon suppliers (Box 5.2).

This conflict between Walmart, the MSC and the Alaskan salmon fishing industry points to a growing tension between a consumer-oriented label and the corporate image of the retailer as the most appropriate assurance for sustainable seafood. Some experts argue that the retailer's corporate image should provide

Box 5.2 Walmart, the MSC and the Alaskan salmon suppliers

Walmart is the largest supermarket company in the world and the dominant retailer in the US. According to the company's sustainability policy, Walmart US 'requires all fresh and frozen, farmed and wild seafood suppliers to become third-party certified as sustainable using Marine Stewardship Council (MSC), Best Aquaculture Practices (BAP) or equivalent standards'.

By June 2012 all uncertified fisheries and aquaculture suppliers had to be actively working towards certification.[19] In 2013–2014 this policy resulted in a conflict with the Alaskan salmon suppliers who, although being one of the first MSC-certified fisheries worldwide in 2000, decided to drop MSC certification because of the costs involved in the certification process. These costs, estimated at US$2.7 million (Foley and Hébert, 2013), were not compensated through higher prices in the market. They resorted to a new programme called Responsible Fisheries Management (RFM), developed by the Alaska Seafood Marketing Institute in collaboration with Global Trust, which was less complicated with lower fees but was not recognised by Walmart as equivalent to MSC certification. This could lead to the exclusion of Alaskan salmon from its stores.

A solution was found through intervention by the Sustainability Consortium at the request of Walmart; it developed eight principles that Walmart can use as a yardstick to evaluate alternative seafood certification programmes and assess their equivalency to the MSC. The primary consideration in the process was to get 'the language right, the transparency and the credibility'. In January 2014 RFM was recognised by Walmart as being equivalent to the MSC, and Alaskan salmon is still available in Walmart.

Sources: www.theguardian.com/sustainable-business/walmart-sustainable-seafood-alaskan-salmon (accessed 19 June 2014); www.nytimes.com/2013/10/22/business/salmon-fishermen-battle-walmart-on-c..... (accessed 28 November 2014); Foley and Hébert (2013)

sufficient trust, whereas sustainable seafood labels only increase costs and confusion. Despite these considerations, consumer-oriented labels still remain the most popular tool to assure sustainability in global seafood provisioning.

Next to importers and retailers, restaurants are key players in seafood GVCs because they handle considerable quantities of seafood and engage directly with consumers. Addressing sustainability considerations when consumers are eating out is far from simple, however (Oosterveer and Spaargaren, 2011). Accessible seafood certification is not well established in restaurants, except some MSC-certified restaurants. In addition, eating out is seen as a moment of luxury and relaxation where complex questions, such as sustainability, are not readily included. Thus some NGOs have developed strategies to influence the procurement strategies of restaurants and food service companies rather than consumers directly.

In summary, retailers have a critical position in the development and application of seafood labels, but they struggle to identify the most appropriate label and how to make the connection with consumer concerns and interests. Retailers also have a precarious position in the seafood movement. On the one hand they are compelled to act in their own commercial interests, but at the same time they must maintain their public image and contribute to building consumer trust. It is exactly this position that makes them so central to the success of the seafood sustainability schemes.

Consumers and labels

The success of private voluntary sustainable seafood labels and certification schemes ultimately depends on their acceptance by consumers. When the number of seafood labels increases, particularly in aquaculture where 'different certification systems make different claims about sustainability, depending on their interpretation of sustainable practices' (Miller and Bush, 2015), consumer support becomes a challenge. The continued proliferation of labels and certification schemes may confuse consumers even when they are concerned about the sustainability of their seafood (Nostvold et al., 2013). Consumers have to decide which (if any) of the labels to trust, and whether they are willing to pay its (surplus) price and make the extra effort of accessing it.

As consumers are not involved directly with a certification scheme themselves, demand has to be created by others, such as NGOs and retailers (Lay, 2012). Engaging consumers in buying certified seafood proves challenging in different parts of the world – for instance in Asia, where more than two-thirds of the world's seafood is consumed but only a few consumers discriminate between labelled and non-labelled products (Boyd and McNevin, 2012). Even in Europe and the US, where many consumers express support for more sustainable seafood provision, their actual buying behaviour often lags behind this verbal commitment. This may be explained as the gap between attitude and behaviour (Padel and Foster, 2005; Vermeir and Verbeke, 2006; Verbeke et al., 2007), as the distinction between the citizen and the consumer roles (Berglund and Matti, 2006; Johnston, 2008), as the perceived imbalance between the higher price of certified seafood and the actual gains (Roheim et al., 2011) or as the relative low importance of environmental concerns compared with taste, freshness, availability and habits (Buergelt et al., 2013). Still others point at the embeddedness of food consumption in highly routinised social practices that do not change swiftly (Spaargaren and Martens, 2005; Jackson, 2006; Shove and Spurling, 2013). These multiple explanations underline that understanding consumers and the changes in their everyday behaviour remains a challenge.

Another fundamental question is whether private labelling essentially means the commodification of seafood sustainability (Luke, 2005; Knowles, 2011; Foley and Hébert, 2013). As labelling and certification schemes seem to be part of market dynamics, their success depends on their uptake in the market. Hence sustainability in seafood provision becomes dependent on the market demand for certified seafood. Market demand depends on consumer buying power, and thus the consumers with the greatest buying power decide on the common definition of 'what is sustainable seafood' and its uptake. The interests and concerns of less affluent consumers are not taken into consideration. Critics thus conclude that sustainability is becoming a commodity bought on the market and defined by the concerns of a select group of consumers. It remains an ongoing challenge for certification to continue to find a balance between commercial interests and the normative goals of the sustainable seafood movement.

Discussion and conclusions

Market-based seafood governance arrangements do not constitute a homogeneous group. They differ with respect to the issues addressed and the ways in which they aim to achieve their goal. What they do have in common is an ambition to shift from producer-driven to a more buyer-driven orientation promoting sustainability in seafood GVCs. However, the extent to which such a model is relevant remains highly questionable given that these initiatives are aimed relatively narrowly at European and US consumers. The debate as to whether the initiatives will have a long-term impact on production is divided into three main arguments.

First, various studies argue that the diversity of claims complicates the ability of consumers to make clear decisions on what is sustainable and what is not, and that the majority of producers around the world (and the majority of production) are unlikely to be included in these initiatives. This has consequences for the long-term impact of these schemes – and highlights the need for them not to overreach their limited transformative capacity. Such a view is increasingly being communicated. For example, the Steering Committee of the State-of-Knowledge Assessment of Standards and Certification (2012) concludes that it is unclear whether aquaculture labels have resulted in significant improvement, and this uncertainty in itself is a driver for further proliferation of new standards that claim to address all impacts instead of focusing on some of the key problems.

To overcome such concerns, we argue that private initiatives need to shift to more integrated approaches, addressing multiple dimensions of sustainability and transforming into what can be termed 'full-chain standards'. They should be oriented not only towards satisfying European and American consumers' concerns, but also those of producers and consumers in other parts of the world.

Second, research has shown that certification has many benefits that are often not recognised. Some argue that while measuring impacts from certification remains challenging because too little attention has been given to the analysis of performance data against the standards that were set, these standards do provide a means of identifying farms that need improvement and provide a detailed roadmap for which changes to make (Accenture, 2009; Tlusty and Tausig, 2014). Others argue that certification should not be seen as a singular solution to transform seafood production, but instead as one tool that can contribute to and stimulate change within a broader regulatory context (e.g. Jonell et al., 2013).

This argument is particularly insightful because it highlights an artificial divide between market-based (or private) forms of governance and state (public) regulation. The classification of (forestry) certification as 'non-state market driven' by Bernstein and Cashore (2007) reflects the now persistent division of these governance approaches. A growing literature, however, has demonstrated how governing seafood and other commodities through the market is closely related to, and often inseparable from, state regulation and governance (e.g. Foley, 2013; Gulbrandsen, 2013). In the context of globalisation, is it unlikely that national governments will replace private initiatives and develop effective transnational regulations for seafood production and trade. Private initiatives are more dynamic and flexible, and their role is probably relevant in the future as well. Nevertheless, governments remain essential in market-based seafood governance initiatives because they create and maintain a legal framework within which private initiatives can emerge and sustain. In many instances governments and private initiatives interact, whereby the latter can 'enhance state capacity by allowing the state to escape innate constraints' (Ponte and Cheyns, 2013: 6).

Third, the market-based instruments outlined above collectively represent a range of instruments in a wider field of governing the transition of the seafood sector to greater sustainability. Debates over whether the proliferation of instruments and claims will lead to inefficiencies in the chain and consumer and producer confusion will likely persist. However, from a governance perspective the proliferation of initiatives can play an essential role in fostering debate and holding not only consumers, producers and governments but also each other to account. Moreover, while certification has become a dominant form of market-based governance, it is by no means the only instrument used in the seafood movement. As we see in the next chapter, the debates that have emerged from the last 20 years of certification have led to a further proliferation of alternative 'beyond labelling' initiatives that respond in large part to the lessons learnt from the experience of recommendation lists and certification schemes.

Notes

1 See http://savedolphins.eii.org/campaigns/dsf/ (accessed 18 May 2018).
2 See https://20.msc.org/what-we-are-doing/our-collective-impact (accessed 18 May 2018).
3 See www.friendofthesea.org (accessed 18 May 2018).
4 See www.seafish.org/rfs/ (accessed 18 May 2018).
5 See https://bapcertification.org/ProgramIntegrity (accessed 18 May 2018).
6 See www.gaalliance.org (accessed 11 June 2014). Interestingly, in December 2014

the GAA signed a memorandum of understanding with a prominent Chinese food service provider, Wuhan Lanesync.

7 GlobalGAP changes its name from EurepGAP in 2007, reflecting the Euro-Retailer Produce Working Group's aspirations to become a global organisation (Henson and Humphrey, 2009).

8 GlobalGAP chief executive officer Kristian Moeller in an interview with Seafood-Source on 2 December 2014. See www.seafoodsource.com/news/aquaculture/globalgap-ceo… (accessed 25 May 2015).

9 See https://aquaculture.ggn.org/en/the-fos-standard.html (accessed 18 May 2018).

10 See www.asc-aqua.org (accessed 11 June 2014).

11 See www.asc-aqua.org/news/certification-update/ (accessed 13 September 2017).

12 See www.msc.org/newsroom/news/western-australian-government-14.5 million-fund-for-fish-and-fisheries (accessed 19 June 2014). 'Western Australian Government announces $14.5 million fund to secure a prosperous future for fish and fisheries.' The fund is split into A$8 million for research and management and A$6.5 million for assessment to gain MSC certification.

13 Only certification through a scheme that fulfils the FAO criteria in its guidelines for the ecolabelling of fish and fishery products from marine-capture fisheries was entitled to such a subsidy (de Vos et al., 2010).

14 See www.cbl.nl/activiteiten/duurzaamheid/duurzame-vis/ (accessed 31 October 2014).

15 See www.nieuws.nl/economie/20111214/Supermarkten-halen-doelstelling-duurzame-vis-niet (accessed 14 November 2014).

16 See note 15. The environmental NGO Stichting Noordzee claims that by now 85 per cent of fresh captured seafood sold through Dutch supermarkets is MSC certified (or equivalent, because there are no good standards available yet for some of the seafood sold). See www.goedevis.nl/Ondernemers/Supermarkt (accessed 31 October 2014).

17 See http://sustainableseafoodcoalition.org/ (accessed 8 August 2017), in particular the SSC Voluntary Code of Conduct on Environmental Claims, August 2014.

18

> The Global Sustainable Seafood Initiative (GSSI) is a unique solution to this problem of multiple certification schemes. For the first time members of the seafood supply chain, NGOs, governmental and intergovernmental organizations and a number of independent experts have come up with a collective, non-competitive approach to provide clarity on seafood certification and ensure consumer confidence in certified seafood.
>
> See www.ourgssi.org/ (accessed 9 August 2017)

19 See http://corporate.walmart.com/global-responsibility/environment-sustainability/sustainable-seafood (accessed 4 July 2014).

References

Ababouch, L., S. Vannuccini and V. Chomo (2016). 'Fisheries, aquaculture utilization and trade: Challenges and opportunities', in B. Onguglo, D. Vivas Eugui and M. Cusi (eds) *Trade and Environment Review 2016: Fish Trade*. Geneva: UNCTAD, 27–34.

Accenture (2009). *Assessment of On-Pack, Wild-Capture Seafood Sustainability Certification Programmes and Seafood Labels*. Zurich: Accenture/WWF International, 138.

Auld, G. and L. H. Gulbrandsen (2010). 'Transparency in nonstate certification: Consequences for accountability and legitimacy', *Global Environmental Politics* 10(3), 97–119.

Bailey, M., S. Bush, P. Oosterveer and L. Larastiti (2016). 'Fishers, Fair Trade, and finding middle ground', *Fisheries Research* 182(Supplement C), 59–68.

Baird, I. G. and N. Quastel (2011). 'Dolphin-safe tuna from California to Thailand: Localisms in environmental certification of global commodity networks', *Annals of the Association of American Geographers* 101(2), 337–355.

Belton, B., D. Little and K. Grady (2009). 'Is responsible aquaculture sustainable aquaculture? WWF and the eco-certification of tilapia', *Society and Natural Resources* 22(9), 840–855.

Belton, B., M. M. Haque, D. C. Little and L. X. Sinh (2011). 'Certifying catfish in Vietnam and Bangladesh: Who will make the grade and will it matter?', *Food Policy* 36(2): 289–299.

Berglund, C. and S. Matti (2006). 'Citizen and consumer: The dual role of individuals in environmental policy', *Environmental Politics* 15(4), 550–571.

Bernstein, S. and B. Cashore (2007). 'Can non-state global governance be legitimate? An analytical framework', *Regulation and Governance* 1(4), 1–25.

Blomquist, J., V. Bartolino and S. Waldo (2015). 'Price premiums for providing eco-labelled seafood: Evidence from MSC-certified cod in Sweden', *Journal of Agricultural Economics*, 66(3), 690–704.

Boström, M. (2006). 'Establishing credibility: Practising standard-setting ideals in a Swedish seafood-labelling case', *Journal of Environmental Policy & Planning* 8(2), 135–158.

Bottema, M., S. Bush and P. Oosterveer (2018). 'Moving beyond the shrimp farm: Spaces of shared environmental risk?', *Geographical Journal* online, DOI https://doi.org/10.1111/geoj.12280.

Boyd, C. E. and A. A. McNevin (2012). 'An early assessment of the effectiveness of aquaculture certification and standards', in Steering Committee of the State-of-Knowledge Assessment of Standards and Certification (ed.) *Toward Sustainability: The Roles and Limitations of Certification*. Washington, DC: RESOLVE Inc., A-35–A-69.

Boyd, C. E., A. A. McNevin, J. Clay and H. M. Johnson (2005). 'Certification issues for some common aquaculture species', *Reviews in Fisheries Science* 13(4), 231–279.

Bronnmann, J. and F. Asche, (2016). 'The value of product attributes, brands and private labels: An analysis of frozen seafood in Germany', *Journal of Agricultural Economics*, 67(1), 231–244.

Brown, J. (2005). 'An account of the dolphin-safe tuna issue in the UK', *Marine Policy* 29(1), 39–46.

Buergelt, D., I. B. Christoph-Schulz, P. Salamon and D. Weible (2013). 'Impacts of sustainability labels on consumers' purchasing decisions for fish', in H. Röcklinsberg and P. Sandin (eds) *The Ethics of Consumption: The Citizen, the Market and the Law*. Wageningen: Wageningen Academic Publishers, 39–45.

Bush, S. R., P. Oosterveer, M. Bailey and A. P. Mol (2015). 'Sustainability governance of chains and networks: A review and future outlook', *Journal of Cleaner Production* 107, 8–19.

Bush, S. R., H. Toonen, P. Oosterveer and A. Mol (2013a). 'The "devils triangle" of MSC certification: Balancing credibility, accessibility and continuos improvement', *Marine Policy* 37, 288–293.

Bush, S. R., B. Belton, D. Hall, P. Vandergeest, F. J. Murray, S. Ponte, P. Oosterveer, M. S. Islam, A. P. J. Mol, M. Hatanaka, F. Kruijssen, T. T. T. Ha, D. C. Little and R. Kusumawati (2013b). 'Certify sustainable aquaculture?', *Science* 341(6150), 1067–1068.

Cambridge, T., S. Martin, F. Nimmo, C. Grieve, S. Walmsley, T. Huntington, R. Cappell and D. Agnew (2011). *Researching the Environmental Impacts of the MSC Certification Programme*. London: MRAG/Poseiden/Meridian Prime.

Darby, M. R. and E. Karni (1973). 'Free competition and the optimal amount of fraud', *Journal of Law and Economics* 16(1), 67–88.

de Krom, M. P. M. M. and A. P. J. Mol (2010). 'Food risks and consumer trust. Avian influenza and the knowing and non-knowing on UK shopping floors', *Appetite* 55(3), 671–678.

de Vos, B. I. and S. R. Bush (2011). 'Far more than market-based: Rethinking the impact of the Dutch Viswijzer (Good Fish Guide) on fisheries' governance', *Sociologia Ruralis* 51(3), 284–303.

de Vos, B., A. Bikker and K. Soma (2010). *Eco-labels voor Visserij en Viskweek. Benchmark aan de Hand van FAO-Richtlijnen*. The Hague: LEI-Wageningen UR.

Deighan, L. K. and L. D. Jenkins (2015). 'Fishing for recognition: Understanding the use of NGO guidelines in fishery improvement projects', *Marine Policy* 51, 476–485.

Driessen, P. P. J., C. Dieperink, F. van Laerhoven, H. A. C. Runhaar and W. J. V. Vermeulen (2012). 'Towards a conceptual framework for the study of shifts in modes of environmental governance – Experiences from the Netherlands', *Environmental Policy and Governance* 22(3), 143–160.

EII (2011). *Earth Island Institute International Tuna Monitoring Program 2011*. Berkeley, CA: Earth Island Institute, 10.

EII (2012). 'Dolphin safe – Consumers', available at www.earthisland.org/dolphin SafeTuna/consumer/ (accessed 1 September 2012).

FAO (2009). *Guidelines for the Ecolabelling of Fish and Fishery Products from Marine Capture Fisheries, Revision 1*. Rome: FAO, 97.

Fiorillo, J. (2014). 'Are the world's retailers and restaurants delivering on their sustainable seafood promises?', *Seafood International (Digital Edition)*, 1–38.

Foley, P. (2013). 'National government responses to Marine Stewardship Council (MSC) fisheries certification: Insights from Atlantic Canada', *New Political Economy* 18(2), 284–307.

Foley, P. and K. Hébert (2013). 'Alternative regimes of transnational environmental certification: Governance, marketization, and place in Alaska's salmon fisheries', *Environment and Planning A* 45(11), 2734–2751.

Gagalyuk, T., J. Hanf and C. Herzlieb (2010). 'Managing supply chains successfully: An empirical testing of success of supply chain networks in the German fish sector', *Acta Agriculturae Scandinavica, Section C – Food Economics* 7(2–4), 139–150.

Greenpeace (2005). *Recipe for Disaster: Supermarkets' Insatiable Appetite for Seafood*. London: Greenpeace UK.

Gudmundsson, E. and C. R. Wessells (2000). 'Ecolabeling seafood for sustainable production: Implications for fisheries management', *Marine Resource Economics* 15(2), 97–113.

Gulbrandsen, L. H. (2009). 'The emergence and effectiveness of the Marine Stewardship Council', *Marine Policy* 33(4), 654–660.

Gulbrandsen, L. H. (2013). 'Dynamic governance interactions: Evolutionary effects of state responses to non-state certification programs', *Regulation and Governance* 8(1), 74–92.

Hall, M. and S. Boyer (1986). 'Incidental mortality of dolphins in the eastern tropical Pacific tuna fishery: Description of a new method and estimation of 1984 mortality', *Report of the International Whaling Commission* 36, 375–381.

Hallstein, E., and S. B. Villas-Boas (2013). 'Can household consumers save the wild fish? Lessons from a sustainable seafood advisory', *Journal of Environmental Economics and Management* 66(1), 52–71.

Hatanaka, M. (2009). 'Certification, partnership, and morality in an organic shrimp network: Rethinking transnational alternative agrifood networks', *World Development* 38(5), 706–716.

Hatanaka, M. (2010). 'Governing sustainability: Examining audits and compliance in a third-party-certified organic shrimp farming project in rural Indonesia', *Local Environment: International Journal of Justice and Sustainability* 15(3), 233–244.

Henson, S., and J. Humphrey (2009). 'The impacts of private food safety standards on the food chain and on public standard-setting processes', paper. Rome: FAO/WHO.

Islam, M. S. (2014). *Confronting the Blue Revolution. Industrial Aquaculture and Sustainability in the Global South.* Toronto, ON: University of Toronto Press.

Jackson, T. (ed.) (2006). *The Earthscan Reader in Sustainable Consumption.* London: Earthscan.

Jacquet, J. L. and D. Pauly (2007). 'The rise of seafood awareness campaigns in an era of collapsing fisheries', *Marine Policy* 31(3), 308–313.

Jacquet, J., J. Hocevar, S. Lai, P. Majluf, N. Pelletier, T. Pitcher, E. Sala, R. Sumaila and D. Pauly (2010). 'Conserving wild fish in a sea of market-based efforts', *Oryx* 44(1), 45–56.

Johnston, J. (2008). 'The citizen-consumer hybrid: Ideological tensions and the case of whole foods market', *Theory and Society* 37(3), 229–270.

Jonell, M., M. Phillips, P. Rönnbäck and M. Troell (2013). 'Eco-certification of farmed seafood: Will it make a difference?', *AMBIO* 42(6), 659–674.

Kelling, I. (2008). '"Responsible retailers": Policy challenges raised by private standards in the seafood sector', in *Achieving a Sustainable Future: Managing Aquaculture, Fishing, Trade and Development.* Nha Trang: International Institute of Fisheries Economics & Trade.

Kemmerly, J. D. and V. Macfarlane (2009). 'The elements of a consumer-based initiative in contributing to positive environmental change: Monterey Bay Aquarium's Seafood Watch Program', *Zoo Biology* 28(5), 398–411.

Kirby, D. S., C. Visser and Q. Hanich (2014). 'Assessment of eco-labelling schemes for Pacific tuna fisheries', *Marine Policy* 43, 132–142.

Knowles, S. (2011). 'The commodification of ethics', in M. Warrier (ed.) *The Politics of Fair Trade. A Survey.* London: Routledge, 24–36.

Lay, K. (2012). 'Seafood ecolabels: For whom and to what purpose?', *Dailhousie Journal of Interdisciplinary Management* 8(2), DOI 10.5931/djim.v8i2.211.

Luke, T. W. (2005). 'Neither sustainable nor development: Reconsidering sustainability in development', *Sustainable Development* 13(4), 228–238.

Manning, S., F. Boons, O. von Hagen and J. Reinecke (2012). 'National contexts matter: The co-evolution of sustainability standards in global value chains', *Ecological Economics* 83, 197–209.

MAP (2017). 'Recommended restaurants and retailers', Mangrove Action Project, available at www.questionyourshrimp.com/sustainable-restaurants/ (accessed 14 November 2018).

Mbengue, M. M. (2011). 'Private standards and WTO law', *BioRes Review* 5(1), 10–11.

Miller, A. M. and S. R. Bush (2015). 'Authority without credibility? Competition and conflict between ecolabels in tuna fisheries', *Journal of Cleaner Production* 107, 137–145.

Moye, P. A. (2010). 'Private certification versus public certification in the international environmental arena: The Marine Stewardship Council and Marine Eco-Label Japan fisheries certification schemes as case studies', *Vanderbilt Journal of Transnational Law* 43, 533–564.

MSC (2011). 'Developing world fisheries participating in the MSC programme', Marine Stewardship Council, London, available at www.msc.org/about-us/credibility/working-with-developing-countries/fisheries-involved-in-the-developing-world-programmes (accessed 23 January 2011).

MSC (2014). MSC Standard Version 2.0, October. London: Marine Stewardship Council.

Nostvold, B. H., I. Kvalvik and J. A. Young (2013). *Sustainable Certification and the Market. Case Studies of the United Kingdom and Sweden*. Tromso: Nofima.

Oosterveer, P. (2006). 'Globalization and sustainable consumption of shrimp: Consumers and governance in the global space of flows', *International Journal of Consumer Studies* 30(5), 465–476.

Oosterveer, P. (2015). 'Authority and legitimacy in governing global food chains', in T. Havinga, F. V. Waarden and D. Casey (eds) *The Changing Landscape of Food Governance. Public and Private Encounters*. Cheltenham and Northampton, MA: Edward Elgar, 117–133.

Oosterveer, P. and G. Spaargaren (2011). 'Organising consumer involvement in the greening of global food flows: The role of environmental NGOs in the case of marine fish', *Environmental Politics* 20(1), 97–114.

Padel, S. and C. Foster (2005). 'Exploring the gap between attitudes and behaviour. Understanding why consumers buy or do not buy organic food', *British Food Journal* 107(8), 606–625.

Parkes, G., J. A. Young, S. F. Walmsley, R. Abel, J. Harman, P. Horvat, A. Lem, A. MacFarlane, M. Mens and C. Nolan (2010). 'Behind the signs – A global review of fish sustainability information schemes', *Reviews in Fisheries Science* 18(4), 344–356.

Ponte, S. (2012). 'The Marine Stewardship Council (MSC) and the making of a market for "sustainable fish"', *Journal of Agrarian Change* 12(2–3), 300–315.

Ponte, S. and E. Cheyns (2013). 'Voluntary standards, expert knowledge and the governance of sustainability networks', *Global Networks* 13(4), 459–477.

Potts, J., A. Wilkings, M. Lynch and S. McFatridge (2016). *State of Sustainability Initiatives Review: Standards and the Blue Economy*. Winnipeg, MB: IISD.

Roheim, C. A. (2009). 'An evaluation of sustainable seafood guides: Implications for environmental groups and the seafood industry', *Marine Resource Economics* 24, 301–310.

Roheim, C. A. and J. G. Sutinen (2006a). *Trade and Market-Related Instruments to Reinforce Fisheries Management Measures to Promote Sustainable Fishing Practices*. Geneva and Paris: International Centre for Trade and Sustainable Development/OECD High Seas Task Force.

Roheim, C. A. and J. G. Sutinen (2006b). 'Trade and marketplace measures to promote sustainable fishing practices', ICTSD Natural Resources. Geneva and Paris: International Centre for Trade and Sustainable Development/OECD High Seas Task Force.

Roheim, C. A., F. Asche and J. I. Santos (2011). 'The elusive price premium for eco-labelled products: Evidence from seafood in the UK market', *Journal of Agricultural Economics* 62(3), 655–668.

Safina, C. (1998). 'What's a fish lover to eat? The *Audubon* guide to seafood', *Audubon* 100, 63–69.

Sampson, G. S., J. N. Sanchirico, C. A. Roheim, S. R. Bush, J. E. Taylor, E. H. Allison, J. L. Anderson, N. C. Ban, R. Fujita, S. Jupiter and J. R. Wilson (2015). 'Secure sustainable seafood from developing countries', *Science* 348(6234), 504–506.

Shove, E. and N. Spurling (2013). *Sustainable Practices. Social Theory and Climate Change*. London and New York: Routledge.

Sogn-Grundvåg, G., T. A. Larsen and J. A. Young (2013). 'The value of line-caught and other attributes: An exploration of price premiums for chilled fish in UK supermarkets', *Marine Policy* 38, 41–44.

Sogn-Grundvåg, G., T. A. Larsen and J. A. Young (2014). 'Product differentiation with credence attributes and private labels: The case of whitefish in UK supermarkets', *Journal of Agricultural Economics* 65(2), 368–382

Spaargaren, G. and S. Martens (2005). 'Globalisation and the role of citizen-consumers in environmental politics', in F. Wijen, K. Zoeteman and J. Pieters (eds) *A Handbook of Globalisation and Environmental Policy. National Government Interventions in a Global Arena*. Cheltenham: Edward Elgar, 211–245.

Standing, A. (2009). *The Growth in Certification of Marine Fisheries in Southern Africa. A Discussion on the Potential Benefits and Challenges*. Pretoria: Institute for Security Studies.

Steering Committee of the State-of-Knowledge Assessment of Standards and Certification (ed.) (2012). *Toward Sustainability: The Roles and Limitations of Certification*. Washington, DC: RESOLVE.

Stemle, A., H. Uchida and C. A. Roheim (2016). 'Have dockside prices improved after MSC certification? Analysis of multiple fisheries', *Fisheries Research* 182, 116–123.

Stolle, D. and M. Micheletti (2013). *Political Consumerism. Global Responsibility in Action*. Cambridge: Cambridge University Press.

Sutton, M. (1998). 'A new paradigm for managing marine fisheries in the next millennium', in D. A. Hancock, D. C. Smith and A. B. Grant, JP (eds) *Developing and Sustaining World Fisheries Resources*, Proceedings of Second World Fisheries Congress. Collingwood, Vic.: CSIRO Publishing, 726–730.

Thrane, M., F. Ziegler and U. Sonesson (2009). 'Eco-labelling of wild-caught seafood products', *Journal of Cleaner Production* 17, 416–423.

Tlusty, M. F. and H. Tausig (2014). 'Reviewing GAA-BAP shrimp farm data to determine whether certification lessens environmental impacts', *Reviews in Aquaculture* 7(2), 107–116.

UNEP (2009). *Certification and Sustainable Fisheries*. Geneva: UNEP.

van Waarden, F. (2012). 'Governing global commons: The public–private protection of fish and forests', in A. Marx, M. Maertens, J. Swinnen and J. Wouters (eds) *Private Standards and Global Governance: Economic, Legal and Political Perspectives*. Cheltenham: Edward Elgar, 15–59.

Vandergeest, P. and A. Unno (2012). 'A new extraterritoriality? Aquaculture certification, sovereignty, and empire', *Political Geography* 31(6), 358–367.

Vasilev, S. (2014). 'How do Dutch retailers and importers of shrimp cope with the multiplicity of certification schemes on sustainable shrimp?', MSc thesis, Wageningen University, Wageningen.

Verbeke, W., F. Vanhonacker, I. Sioen, J. van Camp and S. de Henauw (2007). 'Perceived importance of sustainability and ethics related to fish: A consumer behavior perspective', *AMBIO* 36(7), 580–585.

Vermeir, I. and W. Verbeke (2006). 'Sustainable food consumption: Exploring the consumer "attitude–behavioral intention" gap', *Journal of Agricultural and Environmental Ethics* 19, 169–194.

Wakamatsu, M. and H. Wakamatsu (2017). 'The certification of small-scale fisheries', *Marine Policy* 77, 97–103.
Ward, T. J. and B. Phillips (2008). *Ecolabelling of Seafood: The Basic Concepts*, Chichester: Wiley-Blackwell.
Wessells, C. R. and P. Wallström (1994). 'New dimensions in world fisheries: Implications for US and EC', in G. Anania, C. Carter and A. McCalla (eds) *Trade in Seafood Agricultural Trade Conflicts and GATT: New Dimension in North American–European Agricultural Trade Relations*. Boulder, CO: Westview Press.
Wilkings, A. (2012). 'Fisheries and aquaculture certification: Implications for Southeast Asia', Fisheries Transitions in Southeast Asia, 2011–2014, Working Paper Number 2. Ottawa, ON: University of Ottawa.
Wilkinson, J. (2006). 'Fish: A global value chain driven onto the rocks', *Sociologia Ruralis* 46(2): 139–153.

6 Beyond labelling

Alternative approaches to seafood sustainability

Introduction

Recognising the controversies about and relatively limited impact of certification, as outlined in the previous chapter, different actors in the sustainable seafood movement have experimented with a range of alternative tools and initiatives to stimulate improvements in fisheries' and aquaculture production. This move 'beyond labelling' has been led in large part by two key drivers. First, the perceived failure of the state and mainstream certification schemes (as outlined in the previous two chapters) to push fishers and fish farmers of different sizes and intensities adequately, and across a range of geographies, towards more sustainable production. And second, the expansion of issues that have come to fall under the banner of sustainability in the seafood industry – most commonly beyond environmental performance to take on a range of 'social' issues.

In this chapter we explore a selection of initiatives that attempt to go 'beyond' certification, and in doing so try to understand how they bring together disparate groups of seafood actors to make alternative sustainability claims both in the market and beyond. Building on our understanding of sustainable seafood as a global movement, we question whether the expansion of initiatives and claims creates stronger links between producers and consumers across global space. Does the expansion of the movement to include such alternative initiatives lead to further fragmentation of the seafood movement into a set of market and national/local niches? Or are we instead observing the emergence of globally oriented initiatives that can respond to the demands of more locally embedded values and meanings associated with seafood production and consumption?

We organise our exploration of these initiatives into four areas. First, we identify alternative market initiatives that use a mix of information, equitable distribution and consumer lifestyle claims to incentivise producers to change their production practices. Second, there are advocacy partnerships that, while ostensibly linked to a market logic of changing demand, create frameworks which agitate market actors to bring about change through wider societal awareness raising, shaming and transparency. Third, we look at lifestyle approaches that not only construct meaning around a particular consumer practice but also link the sphere of private life to reflect political views on alternative consumption practices. And fourth, there are resource-based approaches that create market incentives for cooperation between private actors around monitoring, access and control over resources.

'Alternative' market-based approaches

'Market-based' approaches appear at first sight to denote a homogeneous group of governance approaches that appropriate the market logic of supply and demand. Indeed, as outlined in Chapter 5, this rationale has been dominant in what can be considered 'mainstream' market-based sustainable seafood governance arrangements to date. However, the various approaches that we outline in this section exhibit rationales that extend, in different ways, beyond boycotting or buycotting products. Instead, they use value chains as sites of governance that build on but go beyond the market (Bush et al., 2015). In doing so, each of them draws in a wider network of governance actors who use the market as a site of convening and interacting. In this way, it is not only a market logic that emerges but also a range of alternative environmental and developmental logics.

Improvement projects

The notion of improvement has been primarily derived from the systems approach to continual improvement in industrial processes. It's application to fisheries and aquaculture is premised on the need to build the capacity of fishers and farmers to create enough capacity to allow them to comply with the standards of well-established eco-certification schemes such as the MSC and ASC (Deighan and Jenkins, 2015). In order to stimulate fisheries and aquaculture producers that are unable to comply with certification standards, civil society groups have developed what are known as fishery and aquaculture improvement projects (more commonly referred to as FIPs and AIPs; for a conceptualisation of improvement project models and modes of operation see CEA, 2015). Through FIPs and AIPs, NGOs, private companies and fishery consultancy firms support small-scale (developing world) fisheries to move towards more sustainable fishing practices in a systematic, measurable and (ultimately) accountable manner.

FIPs and AIPs build on the kinds of direct technical support that were provided by development aid and NGOs before the rise of the market-oriented sustainable seafood movement. However, they move beyond technical support by also creating partnerships between buyers and those from whom they source their fish to improve their production performance. As argued by Perez-Ramirez et al. (2015), they are commonly seen as a means by which the private sector can assist producers to acquire certification and in a collaborative effort reward fisheries that are not currently able to obtain MSC certification but aspire to do so.

The number of FIPs has increased dramatically since their inception around 2006 (Figure 6.1). Today there are around 90 registered on the website fisheryprogress.org (but numbers of 'unregistered' FIPs are thought to exceed 400 – see Sampson et al., 2015). Examples include the WWF-supported Seafood Savers platform and the broader activities of the Sustainable Fisheries Partnership (SFP), both of which have partnered fishers with retailers in Europe, Australia and the US through FIPs. On this basis, FIPs have created market opportunities for fisheries that are currently not able to obtain MSC certification but nonetheless aspire to get access to Western markets.

At the other end of the value chain major European retailers, including Sainsbury's, Tesco, Ahold Delhaize and Co-op, have engaged in these improvement projects for two reasons: first, the projects match their corporate social responsibility policies – which in the EU are directly steered by NGOs like the WWF (and in the US advised by NGOs like SFP and Ocean Outcomes); and second, and more specifically, these retailers use seafood from FIPs in line with their commitment to sell only MSC sustainable fish (see Iles, 2007; Bush and Roheim, 2018). These retail pledges (discussed later in this chapter) to sell only sustainable (and most commonly MSC-certified) fish were first made in response to NGO pressure targeting their brand credibility value (Roheim et al., 2018). But faced with the realities of limited available certified product, retailers have set up various arrangements for buying and selling 'FIP fish', often marketed under 'sustainable' in-house product lines with marketing slogans claiming commitments to support the development of sustainable seafood.

The rise of improvement projects has been met with both hope and scepticism. On the one hand they provide a pathway for those fisheries not able

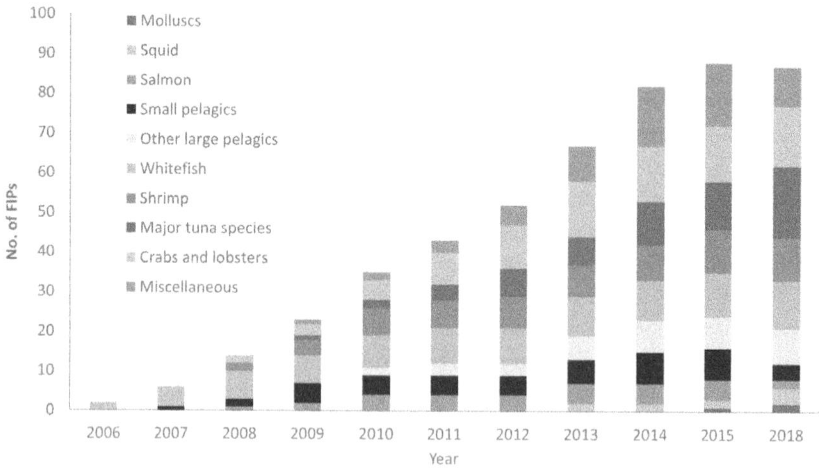

Figure 6.1 Number and distribution of FIPs by species group, 2006–2018.

Source: CEA (2015) updated with data from fisheryprogress.org (2018).

to achieve or demonstrate sustainability immediately to access international markets. But on the other hand they are driving further differentiation of the market for sustainable fish (Bush et al., 2013). Just as certified fisheries are requesting added recognition for going beyond MSC requirements, FIP fisheries are seeking credit for taking a first step on the path to MSC certification. This presents a challenge for those further downstream in the value chain. As Sampson et al. (2015) show, maintaining the supply of this FIP fish appears to outweigh the demands for improvement. If market access is maintained without improvement, only limited 'market-driven' change may be realised.

The rise of national standards

National standards have emerged in both the developed and developing worlds, initiated by states as a means of capturing control over legal compliance in a more inclusive fashion than private standards. But in practice there appear to be different rationales in place for why these national standards have arisen, especially when comparing developed and developing countries (Foley and Hébert, 2013; Samerwong et al., 2018).

In developed countries standards have emerged in an attempt to gain recognition for what states consider to be already robust management systems, and in the same breath reduce their dependency on third-party standard holders. For example, both the government of Iceland and the US state of Alaska have developed a Responsible Fisheries Management (RFM) standard as 'a robust, common sense, practical and cost-effective approach and allows Icelandic fisheries to meet the FAO criteria for credible certification'.[1] The RFM standard was developed in both cases in opposition to the cost and perceived dependency involved in MSC certification. In the case of Alaska, the decision to develop the RFM standard by the Alaska Seafood Marketing Institute was triggered by concerns during the process of MSC recertification about the length and cost of the assessment, coupled with the essentially monopolist position of the MSC and the 'fees' charged for its chain of custody and logo licensing (Foley and Hébert, 2013). But more directly, the institute was concerned that the certification was providing assurance over management activities that were (unusually) required in the constitution of the state. Despite these efforts and the crisis in confidence they caused in the MSC, the Alaskan fishery did complete MSC certification in response to buyer demands.

Similar debates have spread to the EU, again in response to claims around lost sovereign control over assessing the quality of fisheries' management, in addition to cost and the dominant role of MSC. In the EU, which is by far the largest market for sustainably labelled seafood and imports 24 per cent of global traded product (FAO, 2014), considerable attention has been given to its Common Fisheries' Policy. But the debate in the early 2010s turned to the need for a common market for seafood products within which normative goals, such as sustainability, could be more effectively controlled (European

Parliament and Council of the European Union, 2013). In the ensuing debate, some member states opposed to the MSC argued that an EU label would be considered more reliable because of its independence and public legitimacy, thereby reducing the present confusion among consumers and lowering the costs for producers. Other member states argued that yet another label would merely add to the confusing number of existing labels with different claims. While the EU has developed European ecolabels for other consumer items, most notably organic production, it has chosen until now to maintain a third-party regime for seafood. Nevertheless, debate over the position of the MSC and the need for greater control sporadically re-emerges in national parliamentary debates.

In Southeast Asia national standards have also emerged, but it would appear for very different reasons. As an overall trend the emergence of these national standards represents an alternative to the presumption that the greening of production, even beyond fisheries (Schouten and Bitzer, 2015), is now taking place 'from below' rather than as a project of the global North. But there are also indications that the emergence of these standards follows a similar rationale to those that have emerged in the places like the EU and the US: that is, standards that are controlled by the state enable greater sovereign control over resources and people involved in the sector. And as demonstrated in the shrimp aquaculture sector in countries such as Thailand (Samerwong et al., 2017), they also provide a means of ensuring greater inclusion of a wide range of producers in the industry. However, national standards in developing countries have struggled to take hold because buyers appear reluctant to place too much trust in these labels (as in the case of Thailand). This is because the assessment of certification remains with the state, creating an inherent conflict of interest.

Fair Trade fish

Fair Trade certification has been developed as an alternative market-based form of governance that attempts to address social sustainability and encourage environmental protection while redressing power imbalances in international trade and GVCs between consumers in the global North and producers in the global South (Raynolds, 2012; McClenachan et al., 2016). The Fair Trade movement's central aim is to improve the terms of incorporation for smallholder producers selling into the global food market. As such, the US-based Fair Trade Certified (not the Europe-based Fairtrade Foundation) presents a possibility for addressing social sustainability and market inclusion for fishers in developing countries. But while being well established in the market for other commodities, such as tea and coffee (Auld, 2014), the Fair Trade movement has only very recently chosen to engage with fisheries. In 2013 Fair Trade Certified developed its own fisheries' standard (not recognised by Fairtrade International) and certified the Moluccan yellowfin tuna handline fishery in Indonesia in 2014.

A key reason why Fair Trade certification of fisheries has not occurred earlier is the ambiguous role played by traders, who often exploit small-scale fishers through interlocked credit transactions while simultaneously performing crucial social welfare functions (Bush and Oosterveer, 2007; Bailey et al., 2016a). The one and only small-scale coastal community fishery to have successfully applied for Fair Trade Certified managed to rearrange these local relations to the benefit of fishers. But as Bailey et al. (ibid.) argue, the long-term success of the Fair Trade Certified model is dependent on strong working relationships between local intermediaries and processors. Furthermore, relating to experience in other Fair Trade certification sectors (e.g. Raynolds, 2012), it is not yet clear whether the immediate benefits of higher dockside prices and investment in community infrastructure will lead to the empowerment of these communities over the long term.

Although framing social issues in the seafood sector in terms of 'fair trade' opens up space for imagining alternative forms of social and economic development in Southeast Asia, it is questionable whether such approaches will lead to widespread or fundamental changes in how the market value of seafood is distributed. Given the underlying selectivity of ecocertification towards more capital-intensive fisheries and aquaculture systems (Belton et al., 2010; Pérez-Ramírez et al., 2012), it appears unlikely that widespread market access for less powerful players will result. More directed approaches to improved benefit sharing such as Fair Trade Certified hold greater promise in this regard, but widespread upscaling will be required if a significant change is to take place (McClenachan et al., 2016). Furthermore, while Fair Trade Certified is designed for community-level interventions, it remains unclear whether such models can also be applied to improving catch share programmes on industrial vessels where some of the worst forms of human rights abuse are seen.

Organic fisheries and aquaculture

The first organic aquaculture species standard was developed in 1994 and the first national organic aquaculture standard in 2000 in France and the UK, respectively (Prein et al., 2012). Since then organic aquaculture has emerged as one of the fastest-growing food sectors in the world, having expanded 20–30 per cent per year over the last decade (Paul and Vogl, 2012). Since 2000 organic aquaculture has become well established, and in 2008 organic production was over 53,000 MT (value around US$300m), involving 30 species (especially salmon, shrimp and pangasius) in 29 countries (Prein et al., 2012). Nevertheless, overall production is still at a niche scale, with current estimates that only 0.1 per cent of global production is organic. While this is expected to expand to 0.6 per cent by 2030 (Scialabba and Hattam, 2002), the prospects appear to remain limited.

The growth of organic aquaculture has been demand led, with key markets in Europe – Germany, the UK and Switzerland – accounting for the majority of production (Prein et al., 2012). While demand for organic aquaculture products is expected to increase as the middle class in Asia grows and concerns around food safety increase, there are few data to support such claims. Instead what appears to be occurring is mass certification of production systems in countries which already exhibit 'organic' characteristics. As Xie et al. (2013) note, the astronomical growth of organic production in China, from 5,000 MT in 2003 to 85,000 MT in 2012, can in large part be attributed to a combination of domestic and foreign demand. Domestic growth is said to be expanding in response to broader consumer awareness of the food risks that have surfaced in other sectors in China. However, it is apparent that the majority of this growth (more than 90 per cent) came between 2011 and 2012, just as the European Commissioner for Agriculture and Rural Development and the Chinese Minister for Administration of Quality, Supervision, Inspection and Quarantine agreed to open negotiations on a mutual recognition agreement in the field of organic food products. As a consequence, it appears that enormous numbers of predominantly low-input polyculture systems were certified. However, as Xie et al. (ibid.) point out, export to the EU still remains dependent on accreditation by the International Federation of Organic Agricultural Movements, and there is only one standard currently in China, limiting the overall expansion of exports.

It remains to be seen what influence organic labels will have in the seafood industry moving forward. One recent innovation that may increase the market for organic products is the consolidation of national European organic standards into a single EU standard in 2014 (European Union, 2014). This also led to support for the development of an organic aquaculture standard, with the goal of developing a competitive advantage for the European aquaculture industry outside the main species of salmon, bivalves and sea bream. But given the existing low-input aquaculture production systems in other parts of the world, which are in principle organic (although not certified as such), it remains to be seen whether this strategy will provide the long-term advantages for European aquaculture for which it aims.

Traceability

Traceability refers to the structured transfer of information in value chains to address a range of issues related to food safety, and increasingly also the sustainability of products and production practices (Coff et al., 2008; Mol, 2015). Traceability has been given considerable attention in recent years in response to a series of reports relating to high levels of fraud in seafood – in terms of both whether seafood products are sourced from fisheries exhibiting legal fishing practices and whether fish products are being sold under the correct species names (Jacquet and Pauly, 2008; Miller et al., 2012). The assumption underlying traceability as a tool for increasing the sustainability of seafood production and consumption is that greater transparency of information surrounding a product and its origin reduces the risk of IUU fishing. In some cases it may also lead to preferential market access for those fishers and farmers who allow greater oversight of their practices.

The application of traceability to seafood has undergone an extended period of disruption in recent years. Based on various reports of seafood fraud (predominantly involving tuna – see e.g. Stiles et al., 2011), civil society groups have generated considerable attention to the need for traceability. In response, a series of private and state-led initiatives have emerged that collectively present a range of approaches to improving overall transparency in the sector. Miller (2014), for example, identified at least nine consumer-facing traceability systems for tuna alone, either led by NGOs like EcoTrust and Canada's ThisFish or proprietary systems owned and operated by single companies like John West (for an overview of other traceability tools and initiatives see Lewis and Boyle, 2017). While all systems maintained some necessary information, such as where the fish was sourced, they differed considerably in terms of the precision with which this source was communicated and whether this basic information was publicly available. States are also investing in traceability. The US and EU authorities, for instance, have developed catch certification and trade registration requirements to combat IUU fishing. Here states are using chain transparency to feed regulators with information from value-chain actors and practices.

Whether private or public systems will prevail remains to be seen. Given the lack of coordination currently seen across all initiatives, traceability appears to be generating unnecessary informational and capacity burdens for many seafood-exporting countries, which in turn appears to be hindering widespread uptake (Bailey et al., 2016b). Thus, as outlined by Groeneveld et al. (2017), in the long term a series of public–private partnerships may be required to harness effectively the potential role of traceability in governing ocean sustainability.

Whatever system prevails, it is likely that traceability will continue to move to the centre of the sustainable seafood movement, and in doing so will continue to place higher expectations on the overall transparency of the sector. Such expectations are likely to be fuelled by the emerging interest in (and high expectations of) 'blockchain' technologies, with more significant and more secure information sharing in GVCs (see Chow, 2018). But this demand for greater transparency is also likely to generate new tensions between chain actors. On the one hand, greater access to information has the potential to empower seafood producers in the global South, but at the same time it may see greater control over production by those designing and paying for traceability systems in the global North. As outlined by Bailey et al. (2016b), it is yet to be seen whether in its current form traceability will perpetuate as a form of arm's-length monitoring, control and surveillance by large retailers and NGOs, or will rather empower producers to make better decisions over the sustainability of seafood production in response to market demands.

Sustainability partnerships

The role of NGOs in the sustainable seafood movement has changed over time. Whereas early engagement was directly involved in consumer-facing advocacy, later initiatives have become more sophisticated in establishing what has been labelled 'sustainability partnerships' (Bitzer and Glasbergen, 2015) with key actors in global seafood chains. These partnerships work in many ways like the FIPs outlined above. Instead of being oppositional, such as the highly mediagenic campaigns carried out by Greenpeace and the Environmental Defense Fund, some partnerships have created alternative collective-action approaches that support coastal fishing communities while others have sought to raise wider societal awareness. These approaches are not completely disengaged from more oppositional boycott measures, but have supplemented such approaches with what can be labelled conditional boycott measures by providing moral assurance through their own NGO 'brands', by sitting on a board or technical working group or even by co-labelling products with their own logo (Bush and Oosterveer, 2015).

Seafood pledges

Pledges can be understood as voluntary commitments to take action which, although not obligatory, may be experienced as such given broader societal pressure or internal sets of values or norms (Wood, 2006). Corporate sustainable seafood pledges are increasingly made by a wide range of companies which not only aim to increase the number of societal actors who buy and consume more sustainable seafood but also to communicate this clearly to others and convince them to do the same. For companies with a consumer brand, such as Costco, Trader Joe's, Whole Foods, Safeway and Target in the US, Co-op in the UK and the National Retailer Association in the Netherlands, pledges are mainly a means of setting public goals to meet a level of sustainable purchasing – often associated with selling only products that are certified by the MSC or one of the aquaculture standards (for detail see Roheim et al., 2018).

But perhaps more novel is a series of pledges that are being made through the Sustainable Fish Cities campaign, aiming for towns, cities, public institutions and individuals in the UK to 'to buy, serve, eat and promote only sustainable fish' and to share this through social media.[2] This pledge campaign has broadened the focus of sustainability from private sector pledges that are brand- and consumer-oriented to include a larger set of public institutions that procure seafood, such as hospitals, schools and even the 2012 London Olympic Games, that would not otherwise come under consumer-led pressure.[3] Addressing public procurement of seafood, as for most other products, is a widely neglected pressure point for civil society to promote sustainable consumption practices. Although volumes are difficult to come by, clear policies on sustainable fish purchases written for catering services bought by key ministries are seen in a number of OECD countries. In some of these policies there are clear guidelines for MSC-certified or 'equivalent' fish (e.g. those of the Dutch Ministry of Infrastructure and the Environment and the UK Department for Environment, Food and Rural Affairs).[4,5]

Finally, there are various individual pledges made through civil society groups, such as Sustainable Fish Cities, the David Suzuki Foundation and Sea-Choice.[6,7] Their content invariably refers to a direct pledge to purchase and consume sustainable seafood. But by gaining pledges, these platforms are doing more than just enrolling consumers. For example, the David Suzuki Foundation provides a means of engaging with the seafood movement by offering information on a wide range of issues related to seafood sustainability.

Benchmarking

A number of benchmarking initiatives has emerged in response to the perceived complexity and confusion associated with the proliferation of both public and private certification standards. Not all of these initiatives are similar in their goal, approach or organisation. Some are one-off exercises run by NGOs like the WWF (2007), while others, like the Seafood Watch programme, are set up to benchmark their own seafood assessment methodology (Seafood Watch, 2013). Others still, like the GSSI, represent a new model where buyers themselves in partnership with public funding agencies seek clarity over the legitimacy of different standards and their claims around sustainability.

The GSSI is a partnership between retailers, the FAO and the German government's technical development organisation GIZ; it aims to assess, evaluate and ultimately benchmark existing seafood sustainability certification schemes. The GSSI was initially proposed by retailers seeking a method for better evaluation of the multiple seafood standards available to them in the market. By linking to the various normative FAO frameworks, such as the Guidelines for Fisheries Certification and the Code of Conduct for Responsible Fisheries, these retailers and GIZ have seen GSSI as a form of purchasing guidance that draws ultimately on the legitimacy and authority of an international state-based system of food and resource governance.

As outlined by Samerwong et al. (2017), the creation of the GSSI has seen a range of questions emerge around its position and role. Retailers and other industry actors prefer to have minimum acceptable criteria to define a standard that is good enough for GSSI recognition. In contrast, NGOs and standard systems want a higher bar and criteria to differentiate standards and define some standard systems as more effective than others, in the hope of leading to improvement and better practices (cf. Vogel, 2008). However, the diversity of actors involved reveals a range of reasons for participating in the GSSI (Samerwong et al., 2017). Retailers express interest in the (transaction) cost savings involved in purchasing and the reduced need for internal benchmarking exercises. They also argue that costs will be lowered by removing the need for multiple audits associated with certification, as currently required if different buyers request different although similar ecolabels. These notions of cost sharing are supported by suppliers, given the burden they perceive from having to deal with multiple claims and demands around sustainability.

The impact of benchmarking standards is yet to be seen, but the initiative already offers us a series of questions regarding the future of ecocertification. By setting a clear benchmark or bar that is closely linked to access to the US and EU retail markets, the GSSI model may lead to one of two shifts in the certification landscape. First, the benchmark is likely to lead to consolidation of the number of standards that are currently available to producers and buyers alike. Any competition between the standards left would then be likely to be some form of differentiation based on providing new or more stringent claims above the GSSI 'bar'. This would correspond to what Cashore et al. (2007) term a ratcheting-up model of environmental standards, facilitating greater overall improvement. The other transition may be the consolidation of standards at or just over the GSSI bar, representing a set of 'good-enough' standards should buyers want to do away with ecolabels and instead link their brand to sustainability, as suggested by Roheim et al. (2018). One consequence of this retail brand strategy would be a pre-competitive shift in how standards are used. With no need for consumer recognition of labels, the overall shift would then be to cost reduction and, potentially, more public-goods-oriented national standards supported by governments in producer countries.

Pre-competitive collaborations

The drive for direct intervention by civil society groups is also being observed in the activities of the private sector in dealing with what might be termed 'extreme' or critical sustainability events. Pre-competitive collaborations between companies, often with technical support from NGOs, have emerged as one mechanism for achieving industry-wide change on such events or issues. Three examples from the seafood industry are the Seafood Task Force in Thailand, the International Seafood Sustainability Foundation (ISSF) and the Global Salmon Initiative (GSI).

The Seafood Task Force is a collaboration of processing industries, international buyers and supermarkets which came together in response to evidence of 'seafood slavery' centred around the Thai seafood sector. Issues of slavery, or more generally a range of unfree labour practices, emerged through the release of Environmental Justice Foundation (EJF) reports, starting with the publication of *Sold to the Sea* in 2013 (EJF, 2013) and followed by high-profile newspaper articles by Associated Press (2015) and the *Guardian* (Hodal et al., 2014) which led to global coverage. A call for action was made to large Thai seafood processors such as the Charoen Pokphand company, which accounts for 10 per cent of Thai shrimp exports and sells to large retailers in the US and EU like Walmart, Costco and Carrefour – some of which faced legal action from consumer groups in the US for knowingly supplying 'slave seafood' (Marschke and Vandergeest, 2016). The task force, formed in July 2014 in Thailand by seafood producers, major Northern-based retailers, government units and NGOs (including Oxfam, the EJF, the WWF and others), has taken a series of actions with respect to traceability and vessel monitoring, FIPs and lobbying for legal changes to fisheries' laws in Thailand and beyond.

The ISSF is a private conservation organisation established in 2009 by nine of the largest global tuna processors (which together produce 75 per cent of the world's canned tuna) in partnership with WWF US, with the goal of promoting the long-term conservation and sustainable use of tuna stocks, reducing bycatch and promoting ecosystem health (ISSF, 2017). The impetus for forming the ISSF was to provide a platform that could leverage market influence to progress decisions around fishery management by the five tuna RFMOs (see Chapter 4). In addition to providing data and advice on the status of tuna stocks, the ISSF has established voluntary measures, including requirements for members to enrol their vessels in a proactive vessel register that allows vessel owners to demonstrate publicly steps towards upgrading their fishing practices by indicating their progress towards international IUU fishing requirements (see Tolentino-Zondervan et al., 2016). It has also invested in the development of 'ecological' fish attraction devices – floating devices that are deployed by the thousands each year, are associated with high levels of (endangered species) bycatch and can constitute a major source of pollutive marine debris if not recovered by vessels (ISSF, 2014).

The salmon industry has seen the formation of the WWF-facilitated GSI – a coalition between 12 salmon firms across the world representing approximately 55 per cent of global production and the WWF (GSI, 2018a). The GSI (2018b) aims at 'significant reductions in environmental impact and improvements in resource efficiency' of salmon farming, with a focus on biosecurity, feed and nutrition. The explicit goal of the coalition is to work towards ASC compliance. As of October 2016, GSI members reported having approximately 40 per cent of their farms certified (GSI, 2018a).

While different in scope, these three examples of pre-competitive collaboration have a number of things in common. They involve either companies with a controlling share over an industry (as seen with the ISSF over canned tuna and the GSI over salmon aquaculture) or a significant node in global seafood trade flow (as in the case of Thailand). All three examples are characterised by a shared imperative to raise a 'quality baseline' from which the majority of an industry will benefit (Smith, 2008). The companies involved are outwardly trying to go beyond setting standards alone towards shared programmes of research and development to set new industry practices and norms. But while the companies involved can exert a high degree of control, they also hold the potential to gain greater control over an industry or key 'node' in the industry. For instance, some of the founding members of the

ISSF have recently been implicated (and indicted) in a price-fixing scandal in the US.[8]

Lifestyle approaches

Lifestyle approaches to changing consumer lifestyles involve a range of strategies by NGOs and retailers to enrol consumers in a broader range of sustainability-related narratives around seafood. These approaches go beyond more direct forms of advocacy associated with sustainability campaigning, and involve not only constructing meaning around a particular consumer practice but also linking the sphere of private life to reflect political views on alternative consumption practices (Micheletti and Boström, 2014). Traditional areas of lifestyle politics related to seafood have focused on the role of fish in vegetarian diets, or so-called pescatarian diets, reflecting concerns over the sustainability of meat protein consumption more broadly (for more detail see De Backer and Hudders, 2015). But as we outline below, new areas of lifestyle politics related to seafood increasingly go beyond influencing consumer lifestyles to engage the seafood industry in institutional change.

Slow Fish movement

Slow Fish is part of the broader Slow Food movement founded in Italy in 1989 (van Bommel and Spicer, 2011). Central to the philosophy of this movement is the notion that 'with fishing, just as with agriculture, Slow Food strongly believes that every individual can contribute in his or her own small way to changing the mechanisms of a globalized food system based on the intensive exploitation of resources'.[9] Central to this movement is also the 'right to taste' and striking the balance of respect and exchange with nature and the environment while enhancing eating pleasure (Honoré, 2004, cited in Chuenpagdee and Pauly, 2005).The movement works on developing the alternative of local fish as food by raising awareness among consumers, by organizing the biennial Slow Fish fair in Genoa (Italy), by supporting responsible artisanal fishing communities and by organising local initiatives around the world. Many of the activities are oriented to consumers by involving them in 'culinary adventures' and encouraging them to participate in fisheries' resource protection.

Building on a mix of ideas from food sovereignty and local cuisine to notions of fairer trade, Slow Fish is based on three central principles.[10] First, slow fish are 'good' fish – meaning they are 'fresh, delicious and seasonal, satisfying the senses and connected to our culture and local identity'. Second, slow fish are 'clean', 'produced using methods that respect the environment and human health'. And third, slow fish are 'fair', meaning they are sold at accessible prices for consumers but also that fair earnings can 'guarantee decent working and living conditions for small-scale producers and workers'.

By mixing cuisine, sovereignty, environment and health, the Slow Fish movement draws together ideas from what have traditionally been disparate groups in seafood movements. As Chuenpagdee and Pauly (2005) argue, Slow Fish can feed into a wider understanding of slowing down fishing, scaling down fisheries and maintaining small-scale fishing communities. The notion of 'Slow down fishing', they argue, corresponds to a perceived need to reduce 'fast' rates of resource exploitation. 'Scaling down fisheries' refers to a perceived need to reduce fishing capacity through 'appropriate incentives, including a differential tax system favouring the operation of small-scale, owner-operated vessels' (ibid.: 70). Maintaining small-scale fishing communities recognises the importance of local fishing communities in their contributions to food security and viable livelihoods. 'Local' here is underpinned by creating consumer support for marginal small-scale coastal fisheries by building on the idea of moving away from industrialised fisheries, with the expectation of lowering overall fish consumption and returning greater value back to fishing communities. But in doing so, this strategy also emphasises the constriction of species diversity in mainstream markets to 'fewer than 20' and promoting 'lesser-known' and often lower-trophic fish species targeted by the small-scale fishing communities.

There is little hard evidence of the direct impact of the Slow Fish movement. Nevertheless, it does appear to have a discursive impact, and serves as an example for a different way of engaging with seafood production and consumption. In addition, by connecting with other initiatives and movements such as community-supported fisheries (CSFs; see below) and the food sovereignty movement (Sage, 2014), Slow Fish creates a broader impact than its own activities alone.

Celebritisation of seafood

Cooking shows are now broadcast on many television channels around the world. Through these shows and other mass media, chefs have become celebrities and even famous worldwide – an example is Jamie Oliver (Powell and Prasad, 2010; Johnston and Goodman, 2015).

Notably, in the US and the UK celebrity chefs are very visible in the media and known to influence consumer demand (Bowman and Stewart, 2013). Their authority is then used to promote sustainable seafood, for instance through the sale of books that explicitly include such seafood. The cookery books from Hugh Fearnley-Whittingstall clearly illustrate this strategy, as they include many references to the sustainability of the seafood ingredients in his recipes (ibid.). Silver and Hawkins (2017) criticise the role played by these celebrities because they do not really promote the necessary structural changes but only focus on creating a cultural alternative.

Sometimes chefs take a more active stand, and again Hugh Fearnley-Whittingstall is an example: in 2010 he launched the 'Fish Fight' campaign to end the practice of discarding unwanted/unsellable fish in European fisheries.

The Fish Fight campaign made use of multidirectional communication tools (television, newspapers, websites, Facebook, Twitter, YouTube, etc.) and thereby attracted much attention. As outlined by Toonen and Miller (2017), the campaign was supported by 'a range of European celebrities, including other chefs, actors, and sportsmen, as well as some large retailers' and as a result 'more than 870,000 people from 195 countries signed a petition calling for an EU Common Fisheries Policy report and a ban on discarding catch'. The result was impressive. In 2013 the EU banned discarding, and 'while causality cannot be established, it is reported that Fish Fight was a contributing factor behind this decision' (ibid.).

An even more surprising example of a celebrity-centric sustainable seafood campaign was organized by FishLove,[11] a British organisation that advocates for sustainable fishing practices. The non-profit FishLove was set up in 2009 by restaurant owner Nicholas Röhl and actress Greta Scacchi to raise awareness of the unsustainable fishing practices that are destroying the Earth's marine ecosystem. FishLove has run several photo campaigns in which various British celebrities pose nude with dead sea creatures. The funds raised from poster sales are used to support different sustainable fisheries campaigns.

Chefs may be very influential in promoting sustainable seafood, but their activities are mostly occasional and not structural. It thus remains questionable whether such initiatives are able to expand cultural politics of identity into changing consumer practices. As argued by Rojek (2014), such celebrity engagement may, in fact, do little more than reproduce what he calls 'celanthropy', or the transformation of causes into *cause célèbres* via the public involvement of celebrities. The Fish Fight example appears to have moved beyond celanthropy to address more structural policy and political changes to fishing sustainability; the lingering question, however, is how accountable these chefs and their movements are, and what level of representation they can claim in settings like the EU.

Community-supported fisheries

CSFs intend to promote place-based fisheries in a way comparable to community-supported agriculture (Thompson and Coskuner-Balli, 2007). Like the Slow Fish movement, CSFs create an alternative to globalised food provision by organising local food supply to strengthen food sovereignty and reduce dependency on volatile global markets.

Before the start of the fishing season, consumers invest in a share of a CSF's catch. The risks and benefits of this fishing effort are shared by all participants – producers and consumers alike. As such, these initiatives create a support system for producers to improve their economic situation, and maintain local fishing communities and cultures. They also aim to enhance stewardship by directing attention and higher prices to fish caught with sustainable methods, potentially including not just gear and habitat but also broader impacts not usually considered in certification programmes, such as carbon footprints (Olson et al., 2014).

There is no systematic inventory of CSF initiatives, but their number seems to be increasing and the US-based network 'Local Catch' reports 39 CSF members; while not listed, there is a range of CSFs in Europe and other parts of the world (McClenachan et al., 2014).[12] It is too early to conclude whether or not these initiatives are successful, but their advantage is that they extend the direct involvement in implementing sustainability in seafood supply to a much wider group than only the producers themselves (confirming the public dimension of promoting sustainability in seafood governance). The disadvantage of CSFs may be that such initiatives require high levels of commitment from the participants, and therefore their potential to expand beyond selective groups of consumers is likely to be restricted.

These buycott initiatives operate at the high end of the market, and will therefore remain small scale by definition. Still, they illustrate innovative arrangements in the organisation of sustainable (global and local) seafood provision because they actively involve consumers in organising and transforming supply chains. By eliminating the middlemen and retailers, CSFs establish direct relationships between fishers and consumers, create transparency in the supply chain and support local communities.

Resource-based approaches

We group the fourth category of initiatives under the heading of resource-based approaches because they are less oriented to changing seafood supply chains but instead towards creating market incentives for cooperation between private actors around monitoring, access and control over resources (see for instance Grafton et al., 2006). Capture fisheries, as well as most aquaculture farms, operate in wider ecosystems, hence adequate management of these ecosystems is essential for future sustainability. These approaches more explicitly include spatial dynamics, while often being linked more closely with governments through spatial planning, environmental regulation and nature conservation measures. At the same time these approaches pay more attention to collective rather than individual actions.

Entrepreneurial marine protected areas

Entrepreneurial marine protected areas (EMPAs) are networks of small-scale protected areas managed by partnerships between local communities and private operators which 'have a vested economic interest in promoting abundant marine life' (Colwell, 1997: 110). EMPAs are developed within or in combination with state-led marine protected areas, but bring with them the added benefit of protecting discrete pockets of habitat. Private sector involvement is therefore seen as a short-term intervention that can stimulate the development of state-led protected areas by raising local awareness and building local capacity

(Colwell, 1998), a way of providing alternative sources of income to communities in and around these areas, thereby reducing extractive pressure on marine resources (Dixon et al., 1993; Christie and White, 2007), and a long-term means of establishing economic activities around marine conservation that can provide a durable source of funding for protected areas (Bottema and Bush, 2012).

There is a range of EMPA-like initiatives around the world, most of which are located in the global South. Examples include hotels establishing no-take areas in adjacent coastal waters (e.g. Svensson et al., 2009), dive-shop-operated reef conservation (e.g. de Groot and Bush, 2010), public–private partnerships for the conservation of sites with high tourism value (e.g. Teh et al., 2008) and user-fee systems that involve NGOs collecting fees for national or regional governments (e.g. Uyarra et al., 2010).

In a recent review, Bush et al. (2017) argue that the future of EMPAs is dependent on not only the capacity of entrepreneurs to seek and exploit opportunities but also their capacity to consolidate private institutions over the long term. The durability of EMPAs is therefore determined in large part by the support they are able to gain and sustain from the state and local communities. Entrepreneurial activity in marine conservation, involving spatial demarcation of conservation activities, is therefore a highly relational and dynamic process of legitimating private authority. The more institutionalised an arrangement becomes, the more secure the economic and environmental individual and communal payoffs may be, yet concurrently flexibility remains an essential component of entrepreneurialism and innovation. Balancing these juxtaposing elements is a challenge EMPAs face.

It is also evident that state involvement is fundamental in providing the 'action space' for entrepreneurs to exploit and consolidate opportunities while also ensuring they contribute to the stewardship of public resources (Bush et al., 2017). Without a clear framework for state collaboration and support, these private sustainability entrepreneur initiatives are unlikely to be able to establish durable spatially delimited institutions around the conservation of marine resources.

Area-based management

In response to the limitations of farm-level certification, NGOs and the FAO alike have started to move towards area-based management approaches to aquaculture improvement. There are a number of existing concepts describing different configurations for cooperation between aquaculture producers and other actors within a specific geographical area or locality (Brugère et al., 2018). These can be broken down into two general categories: those that enable cooperative forms of production between farmers, and those that go beyond the farm to address shared environmental and production risks.

Farm-focused approaches, including clusters or cooperative forms of production, have been widely promoted in the aquaculture sector as a means of improving the economic performance of smallholders, as well as their ability to participate in GVCs (Kassam et al., 2011). However, contrary to the received wisdom of collective action, which emphasises the willingness and ability of individuals to create favourable group dynamics, the various forms of cooperative production are often conditioned by external support from government, NGOs or the private sector (as seen in the agricultural sector, see Fischer and Qaim, 2012). These cooperative forms of production have also not directly targeted environmental outcomes as a central goal, leaving them focused more on production and development-oriented goals than on environmental sustainability.

Building on ecosystem perspectives on aquaculture, these area-based approaches explicitly aim to address wider environmental risks associated with production across a group of producers (Brugère et al., 2018). The ecosystem approach to aquaculture has been promoted by the FAO as a strategy for developing planning and management frameworks for effectively integrating the aquaculture sector into local (spatial) planning. Central to the approach is the integration of a full range of ecosystem functions and services, improving human well-being and equity, and ensuring that aquaculture is developed in the context of other sectors, policies and goals, as appropriate (Soto et al., 2008). In essence, the FAO approach is focused on states, but the wider notion of area-based management has been taken up by private actors in a model akin to FIPs.

The SFP develops so-called AIPs through an explicit area or zonal approach to management. The aim of this spatially explicit approach is to reduce the cumulative and combined impacts of aquaculture practices, which can be realised through the combined efforts of farmers, governments and other private actors in a given area (SFP, 2018). As variously outlined, this mirrors a more general movement in which private companies are increasingly working beyond farm scale to mitigate risks to productivity that cannot be addressed on-farm by individual farmers (Kissinger et al., 2013; Bottema et al., 2018). Such models also build on the demand of buyers to secure enough seafood for their consumers. These buyers, supported by key NGOs such as the SFP, the WWF and Blue You, are organising these area-based models to include smallholders in spatially explicit 'parks', 'hubs' and 'platforms', with the aim of providing them with services and market access.

Area-based approaches raise the profile of shared risks and take responsibility for solving the problems from farmers alone. But it is unclear how active the private sector can be in promoting such approaches alone. It appears more probable that, as with other spatially explicit sustainability models like EMPAs, area-based aquaculture management will require the ongoing input and support of governments. In what ways such public–private partnerships will develop remains a question as this approach continues to evolve.

Fish banks and environmental credit systems

Environmental credit schemes are being proposed as a means of incentivising environmental reform across a range of sectors, including fisheries (Jack et al., 2008). Environmental credit systems can address productive functions, ecosystem functions or a combination of both (Van Riel et al., 2015). Production-oriented credit systems focus on environmental additions and withdrawals; ecosystem function credits are designed to mitigate critical species and/or habitat loss, or degradation through compensation. Compared to other domains, such as wetlands, the implementation of credit systems in the marine environment is quite limited.

The most elaborated examples are 'incentive-based' fisheries' management proposals, seeking economic reward structures for voluntary changes in fishing practices meeting an agreed-upon management goal (Grafton et al., 2006). Suggested financing mechanisms include Financial Institutions for the Recovery of Marine Ecosystems, environmental certification and variations on more traditional rights-based approaches such as ITQs (e.g. Davies and Rangeley, 2010). Production credit schemes aim to affect the economic efficiency of production to stimulate change. For instance, Kraak et al. (2012) propose an approach to incentivise responsible fishing by providing fishers with a number of fishing-impact credits that they can voluntarily choose to spend, for example by limiting their fishing effort in high-catch or sensitive areas or by fishing longer in lower-catch or less sensitive areas. In their proposals for the reform of the EU Common Fisheries' Policy, the NGOs Marine Conservation Society and ClientEarth (2009) suggested a comparable approach with their fishing credit system – based on a mixed-catch quota, where each species caught in a specified fishing region equals a specific number of 'credits'. Fishers can choose what and how much they catch of each species, as long as the catch does not exceed their total credits allowance.

Proposals for ecosystem function credit schemes include Californian swordfish harvesters mitigating their turtle bycatch by conserving turtle nesting sites (Janisse et al., 2009); compensatory mitigation as an interim management measure for stock recovery (Pascoe et al., 2011); and helping the longline fishing industry to compensate for its seabird bycatch by funding the eradication of rats from ships preying on seabirds at breeding sites (Wilcox and Donlan, 2007). See Box 6.1 for an example.

Fishery credit and banking systems have been criticised for not adequately addressing the root cause of overexploitation. As Van Riel et al. (2015) argue, creating incentives for fishers to decrease their fishing efficiency by changing gear configuration or avoiding vulnerable life stages, species or habitats may also incentivise them to seek out new efficiencies to maintain or regain maximum benefit. It may also be challenging to incentivise resource users to adapt their activities when the management goal is not directly connected to the challenges faced by fishers in multispecies fisheries, freeriding exists, long-term gains for

Box 6.1 Protecting sea turtles with credits

The Endangered Species Act (ESA) in the US includes all species of Pacific sea turtles in its recovery plans. As loggerhead and leatherback sea turtle mortality increased in two Californian drift gillnet fisheries (Julian and Beeson, 1998), the fishing industry implemented ESA's sea turtle conservation management strategy. This led to a significant scaling back of the California driftnet fishery for swordfish through time/area closures. The approach was costly, and ultimately found to be insufficient for the recovery of sea turtle populations because of non-fishery-related impacts on their terrestrial nesting sites.

The California swordfish fishers who use driftnets reasoned that, given the priority for conservation of the species rather than of individual turtles on the high seas, the loss of nesting sites should be a major focus of any coordinated activity. The Federation of Independent Seafood Harvesters, representing the drift netters, subsequently developed a scheme for funding the conservation of Pacific leatherback turtle nesting sites through payments to the Mexican conservation group Asupmatoma, which preserves nesting beaches in Los Cabos. The original plan involved voluntary payments from drift netters, but these were changed to an association-levied 'tax' of a few cents per kilo of swordfish catch to avoid freeriding.

A rough estimate of the conservation effects of Asupmatoma's projects is illustrated by the NGO's statistics of 2007, when 562 nests were protected containing 59,361 eggs, resulting in the release of 41,684 hatched baby sea turtles.

Sources: Van Riel et al. (2015); Janisse et al. (2009)

individual fishers are not evident or fishers fear that conservation measures that hamper their activities will remain in place after the stock has been rebuilt. They are therefore never likely to be a solution in their own right, but instead a mechanisms that can support state regulation.

Direct (digital) enforcement

NGOs are increasingly engaging with new technologies as a means of increasing their advocacy and direct action in seafood governance (Toonen and Bush, 2018). One key player in this new field is Global Fishing Watch, an NGO that brings together Google, Oceana and Skytruth. Established in 2013, Global Fishing Watch aims to make visible all trackable fishing activity in the world's oceans in real time and space using satellite tracking and automatic identification systems for fishing vessels. The stated goals of visualizing these data are to enable the public (and media) to monitor where fishing is occurring, enable the regulators to monitor fishing efforts and enable the fishers to demonstrate that 'they are obeying the law and doing their part'.[13] By bringing together multiple sources of live satellite data with information about a vessel's history with regard to IUU fishing and country of registration, Global Fishing Watch hopes, in an anticipatory way, to alert officials to suspicious vessel movements.

The global aspirations of these tech-based civil-society-led initiatives may be seen as a sort of 'end game' should states ultimately decide to take enforcement action on the basis of the information they receive. But as seen in other sectors employing information-rich technologies to support enforcement, a range of questions remain about such approaches relating to the control and use of the information collected and the willingness of states to take on board such sourced information for enforcement purposes (e.g Gupta, 2008). For example, resistance may come from Southern states around the ownership and control of such information by predominantly Northern nations and initiatives, challenging the neutral 'global' framing of these platforms. Limitations may also be seen in the extent to which some states wish to pursue cases against vessels involved in illegal activities given their own complicit role in sponsoring or owning vessels.

NGOs are also developing the capacity for direct enforcement on the basis of these new surveillance technologies. For instance, the Wildlife Conservation Society in collaboration with the company Conservation Drones is setting up surveillance flights with the Department of Fisheries in Belize to fight IUU fishing.[14] The drone flights take footage of the reefs and coastlines where illegal fishers from nearby Honduras and Guatemala are active; this is analysed and acted upon by government officials. Another example is the work of the EJF in West Africa, where it is using mobile phones and GPS-enabled cameras to empower fishers to identify and report on illegal fishing vessels. By photographing a vessel's call sign, name or unique markings, fishers can send evidence to the EJF, which dispatches vessels to document the incident further or contact enforcement authorities. The EJF also passes on the information to the EU, which logs the information in its wider vessel registers as part of the IUU fishing regulation. These local and regional initiatives seem less problematic for Southern governments as they are more flexible and geared towards collaboration rather than confrontation, unlike many global initiatives.

Conclusions – beyond the mainstream?

The four groups of alternative governance approaches outlined above illustrate the breadth of ongoing innovation in the sustainable seafood movement. What these approaches hold common is the central role of private actors, both companies and NGOs, in developing new ways of incentivising change towards more sustainable seafood production and consumption. But these initiatives go beyond the 'consumption drives demand' approaches of recommendation lists and certification schemes outlined in Chapter 5. Instead, this new array of approaches draws consumers into a wider set of relationships with producers and those trading and selling seafood. They do more than foster price signals along value chains; rather, they foster networked relations in and around global seafood value chains that establish new forms of rule and norm setting, surveillance and agitation that cover production, but also trade and consumption.

Seen as such, these approaches go beyond certification in that they foster what Jacobsen and Dulsrud (2007) refer to as a collection of interactive and iterative processes of social action in markets, governance structures and everyday life. Consumers are enrolled in wider advocacy networks, either directly or immanently, not only to influence demand at the point of sale but also to advocate for changes in the entire system of provision that delivers seafood to market. These approaches can be considered 'beyond certification' strategies, building on existing sustainability standards and expanding them by involving additional actors and dimensions.

Despite the innovation that these approaches represent, they also face several limitations. Many of them have been developed through a sense of urgency for creating radical change in seafood provision and not being satisfied with the progress made by existing labels and government policies. But the actual impact of these alternative approaches remains quite marginal – either because they foster what remain niche food norms (such as fair, organic or inclusive), or they are focused narrowly on inclusiveness of the small-scale sector (e.g. CSFs) or they are deemed too radical for widespread implementation (e.g. credit schemes). In spite of these approaches being established for the benefit of small-scale and/or developing world fisheries, they also appear limited to the global North rather than being initiatives in and of the global South.

Yet despite these limitations, the approaches outlined here have opened up the space for reflection and experimentation on how private actors, often in step with states, can foster more sustainable seafood production. As such, they open up innovative ways for setting a sustainability agenda that is more inclusive and therefore potentially more effective for driving a genuinely global seafood movement. As we discuss in the following chapter, they may provide a basis on which to create new geographies of sustainable seafood production and consumption.

Notes

1 See www.responsiblefisheries.is/certification/ (accessed 30 May 2018).
2 See www.sustainweb.org/sustainablefishcity/ (accessed 10 October 2017).
3 One example is the Sustainable Fish Cities pledge taken by Brakes, the UK's largest wholesale foodservice supplier; see www.sustainweb.org/news/brakes/ (accessed 17 October 2017).
4 See www.gov.uk/government/uploads/system/uploads/attachment_data/file/419245/balanced-scorecard-annotated-march2015.pdf (accessed 11 October 2017).
5 See http://english.rvo.nl/sites/default/files/2013/12/Criteriadocument%20Catering.pdf (accessed 11 October 2017).
6 See www.sustainweb.org/sustainablefishcity/ (accessed 11 October 2017).
7 See http://action.davidsuzuki.org/seafood-pledge (accessed 11 October 2017).
8 See for example http://money.cnn.com/2018/05/16/news/companies/bumble-bee-foods-christopher-lischewski/index.html (accessed 16 November 2018) and www.undercurrentnews.com/2018/05/23/after-walmart-chicken-of-the-sea-says-more-settlements-could-follow/ (accessed 16 November 2018).

9 See http://slowfood.com/slowfish/pagine/eng/pagina.lasso?-id_pg=40 (accessed 4 June 2018).
10 See http://slowfood.com/slowfish/pagine/eng/pagina.lasso?-id_pg=56 (accessed 4 June 2018).
11 See https://fishlove.co.uk/ (accessed 7 November 2015).
12 See www.localcatch.org (accessed 7 May 2015).
13 See http://globalfishingwatch.org/ (accessed 11 October 2017).
14 See http://news.nationalgeographic.com/news/2014/07/140718-drones-illegal-fishing-pirate-belize-ocean/ (accessed 11 October 2017).

References

Auld, G. (2014). *Constructing Private Governance: The Rise and Evolution of Forest, Coffee, and Fisheries Certification*. New Haven, CT: Yale University Press.

Associated Press (2015) 'An AP investigation helps free slaves in the 21st century', Associated Press, available at www.ap.org/explore/seafood-from-slaves/ (accessed 16 November 2018).

Bailey, M., S. Bush, P. Oosterveer and L. Larastiti (2016a). 'Fishers, Fair Trade, and finding middle ground', *Fisheries Research* 182(Supplement C), 59–68.

Bailey, M., S. R. Bush, A. Miller and M. Kochen (2016b). 'The role of traceability in transforming seafood governance in the global South', *Current Opinion in Environmental Sustainability* 18, 25–32.

Belton, B., F. Murray, J. Young, T. Telfer and D. C. Little (2010). 'Passing the Panda standard: A TAD off the mark?', *AMBIO* 39(1), 2–13.

Bitzer, V. and P. Glasbergen (2015). 'Business–NGO partnerships in global value chains: Part of the solution or part of the problem of sustainable change?', *Current Opinion in Environmental Sustainability* 12, 35–40.

Bottema, M. J. M. and S. R. Bush (2012). 'The durability of private sector-led marine conservation: A case study of two entrepreneurial marine protected areas in Indonesia', *Ocean & Coastal Management* 61(Supplement C), 38–48.

Bottema, M., S. Bush and P. Oosterveer (2018). 'Moving beyond the shrimp farm: Spaces of shared environmental risk?', *Geographical Journal* online, DOI https://doi.org/10.1111/geoj.12280.

Bowman, P. and B. D. Stewart (2013). 'Celebrity chefs and the sustainable seafood movement: Smokescreens or a dish to savour?', Marine Ecosystem Management Report No. 4. York: University of York.

Brugère, C., J. Aguilar-Manjarrez, M. C. Beveridge and D. Soto (2018). 'The ecosystem approach to aquaculture 10 years on – A critical review and consideration of its future role in blue growth', *Reviews in Aquaculture* online, DOI https://doi.org/10.1111/raq.12242.

Bush, S. R. and P. Oosterveer (2007). 'The missing link: Intersecting governance and trade in the space of place and the space of flows', *Sociologia Ruralis* 47(4), 384–399.

Bush, S. R. and P. Oosterveer (2015). 'Vertically differentiating environmental standards: The case of the Marine Stewardship Council', *Sustainability* 7(2), 1861–1883.

Bush, S. R. and C. A. Roheim (2018). 'The shifting politics of sustainable seafood consumerism', in M. Boström, M. Micheletti and P. Oosterveer (eds) *The Oxford Handbook of Political Consumerism*. Oxford: Oxford University Press, DOI https://doi.org/10.1093/oxfordhb/9780190629038.013.16.

Bush, S. R., M. Bottema and J. J. Midavaine (2017). 'Sustainability entrepreneurship in marine protected areas', in K. Nicolopoulou, M. Karatas-Ozkan, F. Janssen and J. M. Jermier (eds) *Sustainable Entrepreneurship and Social Innovation*. London: Routledge, 144–160.

Bush, S. R., P. Oosterveer, M. Bailey and A. P. Mol (2015). 'Sustainability governance of chains and networks: A review and future outlook', *Journal of Cleaner Production* 107, 8–19.

Bush, S. R., H. Toonen, P. Oosterveer and A. P. Mol (2013). 'The "devils triangle" of MSC certification: Balancing credibility, accessibility and continuous improvement', *Marine Policy* 37, 288–293.

Cashore, B., G. Auld, S. Bernstein and C. McDermott (2007). 'Can non-state governance "ratchet up" global environmental standards? Lessons from the forest sector', *Review of European Community & International Environmental Law* 16(2), 158–172.

CEA (2015). 'Summary findings from the global landscape review of fishery improvement projects (FIPs).' San Francisco, CA: California Environmental Associates.

Chow, C. (2018). 'Blockchain for good? Improving supply chain transparency and human rights management', *Governance Directions* 70(1), 39.

Christie, P. and A. T. White (2007). 'Best practices for improved governance of coral reef marine protected areas', *Coral Reefs* 26(4), 1047–1056.

Chuenpagdee, R. and D. Pauly (2005). 'Slow Fish: Creating new metaphors for sustainability', *FAO Fisheries Reports* 782, 69.

Coff, C., M. Korthals and D. Barling (2008). *Ethical Traceability and Communicating Food*. Dordrecht: Springer.

Colwell, S. (1997). 'Entrepreneurial marine protected areas: Small-scale commercially supported coral reef protected areas', in M. E. Hatziolos, A. J. Hooten and M. Fodor (eds) *Coral Reefs: Challenges and Opportunities for Sustainable Management*. Washington, DC: World Bank, 110–114.

Colwell, S. (1998). 'Dive-tourism and private stewardship of small-scale coral reef marine protected areas', paper presented at International Tropical Marine Ecosystems Management Symposium, Townsville, Great Barrier Reef Marine Park Authority, available at http://msm-host3-reefed.gbrmpa.gov.au/__data/assets/pdf_file/0007/1879/itmems_217-221_s05_2.pdf (accessed 22 November 2018).

Davies, R. W. D. and R. Rangeley (2010). 'Banking on cod: Exploring economic incentives for recovering Grand Banks and North Sea cod fisheries', *Marine Policy* 34(1), 92–98.

De Backer, C. J. S. and L. Hudders (2015). 'Meat morals: Relationship between meat consumption, consumer attitudes towards human and animal welfare and moral behavior', *Meat Science* 99, 68–74.

de Groot, J. and S. R. Bush (2010). 'The potential for dive tourism led entrepreneurial marine protected areas in Curacao', *Marine Policy* 34(5), 1051–1059.

Deighan, L. K. and L. Jenkins (2015). 'Fishing for recognition: Understanding the use of NGO guidelines in fishery improvement projects', *Marine Policy* 51, 476–485.

Dixon, J. A., L. F. Scura and T. van't Hof (1993). 'Meeting ecological and economic goals: Marine parks in the Caribbean.' Washington, DC: World Bank, 23.

EJF (2013). *Sold to the Sea – Human Trafficking in Thailand's Fishing Industry*. London: Environmental Justice Foundation.

European Parliament and Council of the European Union (2013). 'Regulation (EU) No. 1379/2013 of the European Parliament and of the Council.' Brussels: European Union.

European Union (2014). 'Action plan for the future of organic production in the European Union.' Brussels: European Commission.

FAO (2014). *The State of World Fisheries and Aquaculture 2014*. Rome: FAO.

Fischer, E. and M. Qaim (2012). 'Linking smallholders to markets: Determinants and impacts of farmer collective action in Kenya', *World Development* 40(6), 1255–1268.

Foley, P. and K. Hébert (2013). 'Alternative regimes of transnational environmental certification: Governance, marketization, and place in Alaska's salmon fisheries', *Environment and Planning A* 45(11), 2734–2751.

Grafton, Q. R., R. Arnason, T. Bjørndal, D. Campbell, H. F. Campbell, C. W. Clark, R. Connor, D. P. Dupont, R. Hannesson, R. Hilborn, J. E. Kirkley, T. Kompas, D. E. Lane, G. R. Munro, S. Pascoe, D. Squires, S. I. Steinshamn, B. R. Turris and Q. Weninger (2006). 'Incentive-based approaches to sustainable fisheries', *Canadian Journal of Fisheries and Aquatic Sciences* 63(3), 699–710.

Groeneveld, R., S. R. Bush and M. Bailey (2017). 'Private governance of ocean resources', in P. A. L. D. Nunes, L. E. Svensson and A. Markandya (eds) *Handbook on the Economics and Management of Sustainable Oceans*. Nairobi: UNEP, 416–428.

GSI (2018a) 'The path to the future we all want starts with transparency', Global Salmon Initiative, Brussels, 26 April, available at https://globalsalmoninitiative.org/en/news/the-path-to-the-future-we-all-want-starts-with-transparency/ (accessed 16 November 2018).

GSI (2018b) 'The future of aquaculture', Global Salmon Initiative, available at https://globalsalmoninitiative.org/en/what-is-the-gsi-working-on/the-future-of-aquaculture/ (accessed 16 November 2018).

Gupta, A. (2008). 'Transparency under scrutiny: Information disclosure in global environmental governance', *Global Environmental Politics* 8(2), 1–7.

Hodal, K., C. Kelly and F. Lawrence (2014) 'Revealed: Asian slave labour producing prawns for supermarkets in US, UK', *Guardian*, 10 June, available at www.theguardian.com/global-development/2014/jun/10/supermarket-prawns-thailand-produced-slave-labour (accessed 16 November 2018).

Iles, A. (2007). 'Making the seafood industry more sustainable: Creating production chain transparency and accountability', *Journal of Cleaner Production* 15(6), 577–589.

ISSF (2014). 'Report of the ISSF workshop on FADs as ecological traps', ISSF 2014-03, International Seafood Sustainability Foundation, available at https://iss-foundation.org/download-monitor-demo/download-info/issf-technical-report-2014-03-report-of-the-issf-workshop-on-fads-as-ecological-traps/ (accessed 16 November 2018).

ISSF (2017). 'About ISSF: Enabling sustainability through science and collaboration', available at http://iss-foundation.org/who-we-are/about/ (accessed 24 April 2017).

Jack, B. K., C. Kousky and K. R. E. Sims (2008). 'Designing payments for ecosystem services: Lessons from previous experience with incentive-based mechanisms', *Proceedings of the National Academy of Sciences* 105(28), 9465–9470.

Jacobsen, E. and A. Dulsrud (2007). 'Will consumers save the world? The framing of political consumerism', *Journal of Agricultural and Environmental Ethics* 20(5), 469–482.

Jacquet, J. L. and D. Pauly (2008). 'Trade secrets: Renaming and mislabeling of seafood', *Marine Policy* 32(3), 309–318.

Janisse, C., D. Squires, J. A. Seminoff and P. H. Dutton (2009). Conservation investments and mitigation: The California drift gillnet fishery and Pacific sea turtles', in

R. Hilborn, D. Squires, M. Williams, M. Tait and R. Q. Grafton (eds) *Handbook of Marine Fisheries Conservation and Management*. New York: Oxford University Press, 231–240.

Johnston, J. and M. K. Goodman (2015). 'Spectacular foodscapes', *Food, Culture & Society* 18(2), 205–222.

Julian, F. and M. Beeson (1998). 'Estimates of marine mammal, mammal, turtle, and seabird mortality for two California gillnet fisheries', *Fisheries Bulletin* 96(2), 271–284.

Kassam, L., R. Subasinghe and M. Phillips (2011). 'Aquaculture farmer organizations and cluster management – Concepts and experiences', Fisheries and Aquaculture Technical Paper No. 563. Rome: FAO.

Kissinger, G., A. Brasser and L. Gross (2013). 'Reducing risk: Landscape approaches to sustainable sourcing', report. Washington, DC: Landscapes for People, Food and Nature Initiative.

Kraak, S. B. M., D. G. Reid, H. D. Gerritsen, C. J. Kelly, M. Fitzpatrick, E. A. Codling and E. Rogan (2012). '21st century fisheries management: A spatio-temporally explicit tariff-based approach combining multiple drivers and incentivising responsible fishing', *ICES Journal of Marine Science* 69(4), 590–601.

Lewis, S. G. and M. Boyle (2017). 'The expanding role of traceability in seafood: Tools and key initiatives', *Journal of Food Science* 82(S1), A13–A21.

Marine Conservation Society and ClientEarth (2009). 'CFP reform proposal. The fishing credits system.' London: Marine Conservation Society/ClientEarth.

Marschke, M. and P. Vandergeest (2016). 'Slavery scandals: Unpacking labour challenges and policy responses within the off-shore fisheries sector', *Marine Policy* 68, 39–46.

McClenachan, L., S. T. M. Dissanayake and X. Chen (2016). 'Fair Trade fish: Consumer support for broader seafood sustainability', *Fish and Fisheries* 17(3), 825–838.

McClenachan, L., B. P. Neal, D. Al-Abdulrazzak, T. Witkin, K. Fisher and J. N. Kittinger (2014). 'Do community supported fisheries (CSFs) improve sustainability?', *Fisheries Research* 157, 62–69.

Micheletti, M. and M. Boström (2014). 'Political consumerism: Consumer choice, information, and labeling', in P. B. Thompson and D. M. Kaplan (eds) *Encyclopedia of Food and Agricultural Ethics*. Dordrecht: Springer, 1508–1515.

Miller, A. M. (2014). 'Governance innovation networks for sustainable tuna', PhD thesis, Wageningen University, Wageningen.

Miller, D., A. Jessel and S. Mariani (2012). 'Seafood mislabelling: Comparisons of two Western European case studies assist in defining influencing factors, mechanisms and motives', *Fish and Fisheries* 13(3), 345–358.

Mol, A. P. J. (2015). 'Transparency and value chain sustainability', *Journal of Cleaner Production* 107(Supplement C), 154–161.

Olson, J., P. M. Clay and P. Pinto da Silva (2014). 'Putting the seafood in sustainable food systems', *Marine Policy* 43, 104–111.

Pascoe, S., C. Wilcox and C. J. Donlan (2011). 'Biodiversity offsets: A cost-effective interim solution to seabird bycatch in fisheries?', *PLoS One* 6(10), e25762.

Paul, B. G. and C. R. Vogl (2012). 'Key performance characteristics of organic shrimp aquaculture in southwest Bangladesh', *Sustainability* 4(5), 995–1012.

Pérez-Ramírez, M., M. Castrejón, N. L. Gutiérrez and O. Defeo (2015). 'The Marine Stewardship Council certification in Latin America and the Caribbean: A review of experiences, potentials and pitfalls', *Fisheries Research* 182, 50–58.

Pérez-Ramírez, M., B. Phillips, D. Lluch-Belda and S. Lluch-Cota (2012). 'Perspectives for implementing fisheries certification in developing countries', *Marine Policy* 36(1), 297–302.

Powell, H. and S. Prasad (2010). '"As seen on TV": The celebrity expert. How taste is shaped by lifestyle media', *Cultural Politics* 6(1), 111–124.

Prein, M., S. Bergleiter, M. Ballauf, D. Brister, M. Halwart, K. Hongrat, J. Kahle, T. Lasner, A. Lem, O. Lev, C. Morrison, Z. Shehadeh, A. Stamer and A. A. Wainwright (2012). 'Organic aquaculture: The future of expanding niche markets', in R. P. Subasinghe, J. R. Arthur, D. M. Bartley, S. S. De Silva, M. Halwart, N. Hishamunda, C. V. Mohan and P. Sorgeloos (eds) *Farming the Waters for People and Food*, proceedings of Global Conference on Aquaculture, Phuket. Rome and Bangkok: FAO/NACA, 549–568.

Raynolds, L. T. (2012). 'Fair Trade: Social regulation in global food markets', *Journal of Rural Studies* 28(3), 276–287.

Roheim, C., S. R. Bush, J. N. Sanchirico, F. Asche and H. Uchida (2018). 'The evolution and future of the sustainable seafood market', *Nature Sustainability* 1, 392–398.

Rojek, C. (2014). '"Big Citizen" celanthropy and its discontents', *International Journal of Cultural Studies* 17(2), 127–141.

Sage, C. (2014). 'The transition movement and food sovereignty: From local resilience to global engagement in food system transformation', *Journal of Consumer Culture* 14(2), 254–275.

Samerwong, P., S. R. Bush and P. Oosterveer (2017). 'Metagoverning aquaculture standards: A comparison of the GSSI, the ASEAN GAP, and the ISEAL', *Journal of Environment & Development* 26(4), 429–451.

Samerwong, P., S. R. Bush and P. Oosterveer (2018). 'Implications of multiple national certification schemes for shrimp aquaculture in Thailand', *Aquaculture* 493, 319–327.

Sampson, G. S., J. N. Sanchirico, C. A. Roheim, S. R. Bush, J. E. Taylor, E. H. Allison, J. L. Anderson, N. C. Ban, R. Fujita and S. Jupiter (2015). 'Secure sustainable seafood from developing countries', *Science* 348(6234), 504–506.

Schouten, G. and V. Bitzer (2015). 'The emergence of Southern standards in agricultural value chains: A new trend in sustainability governance?', *Ecological Economics* 120, 175–184.

Scialabba, N. and C. Hattam (2002). *Organic Agriculture, Environment and Food Security*. Rome: FAO.

Seafood Watch (2013). 'Eco-certification benchmarking project', Monterey Bay Aquarium Seafood Watch, Monterey, CA, available at www.seafoodwatch.org/-/m/sfw/pdf/eco-certifications/reports/mba-seafoodwatch-benchmarking-faqs.pdf?la=en (accessed 16 November 2018).

SFP (2018). 'Aquaculture improvement projects', available at www.sustainablefish.org/Programs/Aquaculture/Aquaculture-Improvement-Projects (accessed 4 June 2018).

Silver, J. J. and R. Hawkins (2017). '"I'm not trying to save fish, I'm trying to save dinner": Media, celebrity and sustainable seafood as a solution to environmental limits', *Geoforum* 84, 218–227.

Smith, B. G. (2008). 'Developing sustainable food supply chains', *Philosophical Transactions of the Royal Society B: Biological Sciences* 363(1492), 849–861.

Soto, D., J. Aguilar-Manjarrez and N. Hishamunda (2008). 'Building an ecosystem approach to aquaculture', paper presented at FAO/Universitat de les Illes Balears expert workshop, Palma de Mallorca, 7–11 May 2007, FAO Fisheries and Aquaculture Proceedings No. 14. Rome: FAO.

Stiles, M. L., H. Lahr, W. Lahey, E. Shaftel, D. Bethel, J. Falls and M. F. Hirshfield (2011). 'Bait and switch: How seafood fraud hurts our oceans, our wallets and our health', report. Washington, DC: Oceana.

Svensson, P., L. D. Rodwell and M. J. Attrill (2009). 'Privately managed marine reserves as a mechanism for the conservation of coral reef ecosystems: A case study from Vietnam', *AMBIO* 38(2), 72–78.

Teh, L. C., L. S. Teh and F. C. Chung (2008). 'A private management approach to coral reef conservation in Sabah, Malaysia', *Biodiversity and Conservation* 17(13), 3061–3077.

Thompson, C. J. and G. Coskuner-Balli (2007). 'Enchanting ethical consumerism', *Journal of Consumer Culture* 7(3), 275–303.

Tolentino-Zondervan, F., P. Berentsen, S. Bush, J. Idemne, R. Babaran and A. O. Lansink (2016). 'Comparison of private incentive mechanisms for improving sustainability of Filipino tuna fisheries', *World Development* 83, 264–279.

Toonen, H. M. and S. R. Bush (2018). 'The digital frontiers of fisheries governance: Fish attraction devices, drones and satellites', *Journal of Environmental Policy & Planning* online, DOI https://doi.org/10.1080/1523908X.2018.1461084.

Toonen, H. M. and A. M. M. Miller (2017). 'Social media and celebrity chefs: Promoting sustainability in fisheries', *Solutions Journal* 8(3), available at www.thesolutionsjournal.com/article/social-media-celebrity-chefs-promoting-sustainability-fisheries/ (accessed 16 November 2018).

Uyarra, M. C., J. A. Gill and I. M. Côté (2010). 'Charging for nature: Marine park fees and management from a user perspective', *AMBIO* 39(7), 515–523.

van Bommel, K. and A. Spicer (2011). 'Hail the snail: Hegemonic struggles in the slow food movement', *Organization Studies* 32(12), 1717–1744.

Van Riel, M. C., S. R. Bush, P. A. M. van Zwieten and A. P. J. Mol (2015). 'Understanding fisheries credit systems: Potentials and pitfalls of managing catch efficiency', *Fish and Fisheries* 16(3), 453–470.

Vogel, D. (2008). 'Private global business regulation', *Annual Review of Political Science* 11, 261–282.

Wilcox, C. and C. J. Donlan (2007). 'Compensatory mitigation as a solution to fisheries bycatch – Biodiversity conservation conflicts', *Frontiers in Ecology and the Environment* 5(6), 325–331.

Wood, S. (2006). 'Voluntary environmental codes and sustainability', in Benjamin J. Richardson and Stepan Wood (eds) *Environmental Law for Sustainability*. Oxford: Hart Publishing, 229–276.

WWF (2007). 'Benchmarking study on international aquaculture certification programmes', World Wildlife Fund, available at wwf.panda.org/wwf_news/?119260/Benchmarking-Study-Certification-Programmes-for-Aquaculture (accessed 16 November 2011).

Xie, B., J. Qin, H. Yang, X. Wang, Y.-H. Wang and T.-Y. Li (2013). 'Organic aquaculture in China: A review from a global perspective', *Aquaculture* 414/415, 243–253.

Part III

The future of sustainable seafood

7 Towards a social understanding of the sustainable seafood movement

Introduction

Nearly all the initiatives reviewed in the preceding chapters are driven by theories of change that emphasise a reductionist market logic focused on information asymmetries and price signals in a stylised GVC that distributes costs and benefits efficiently (Gudmundsson and Wessells, 2000; Crona et al., 2016b). They are also based on ambitions to improve methodologies for assessing the performance of fishers and fish farmers, or make the seafood industry more fair and transparent. Following Mutersbaugh (2005), what emerges is an obsession (or 'fetishisation') with how to assess sustainability that clouds rather than enlightens the relations of assessment and production that enable or limit producers in making sustainability improvements.

In this chapter we challenge this reductionist perspective by arguing that the sustainable seafood movement is a fundamentally social phenomenon that functions beyond the narrow confines of the market. Seen as such, the various governance tools outlined in the preceding three chapters are social, in that they mediate the networks of NGOs, policymakers and industry partners which structure flows of information, moral authority and value exchange across global space, and which in turn (re)shape the performance of sustainable seafood.

The chapter is divided into two parts. First, we build on the theory reviewed in Chapter 2 to explore the social dynamics of the global sustainable seafood movement. Here we unpack GVCs into their constitutive global flows and practices by 'looking behind the market stage' (Platteau, 1994) to understand how economic practices of production, trade and consumption relate with socially negotiated rules, norms and practices. Second, we use this social understanding of the sustainable seafood movement to outline a new theoretical perspective for understanding environmental movements, by focusing on how practices from production through to consumption are shaped by multiple social and material flows. This framework provides a starting point for understanding how markets and governance operate in the global seafood industry.

Revisiting the social geographies of sustainable seafood

The following subsections reflect on the dynamics of the global seafood movement outlined in Chapter 3 in light of the governance arrangements reviewed in Chapters 4–6. We revisit how global/local dynamics, inclusion/exclusion, public/private and convergence/divergence aspects affect who can and cannot contribute to the resolution of sustainability issues. By overlaying these dynamics, we outline the need for a more comprehensive (and therefore social) understanding of trade and consumption.

Global–local dynamics

The initiatives reviewed in this book illustrate how the seafood movement operates at the interface between the global and the local. On the one hand the movement is promoting seafood sustainability as a universal imperative, while on the other hand the various initiatives struggle with the realisation that production practices cannot be standardised to a single level of performance. The result is the emergence of a series of what Tsing (2011) labelled 'frictions' between global and local norms, practices and institutions. But what also emerges from the preceding chapters is that the sustainable seafood movement has responded to such frictions by the further development of governance tools and approaches.

The early phase of the movement focused on rolling out a global agenda that assumed moral authority over particular local interests. Inherent to this agenda was the perceived failure of especially developing countries (both states and producers) to guarantee safe and (subsequently) sustainable products for consumption in the major 'developed' markets of the US and EU. This led to the emergence of global boycott and buycott tools and arrangements outlined in Chapter 5, nearly all of which were (and continue to be) based on the moral authority of a singular global imperative for sustainability of 'shared' ocean resources. While a degree of variation is built into evaluation methodologies for smallholders or specific fisheries and aquaculture species, a homogenising assumption of production has to a large extent been maintained. The consequence, as seen in the global sustainability movements related to coffee, tea and timber (Ponte, 2002; Taylor et al., 2005; Dolan, 2010; Neilson, 2014), is that the effectiveness and inclusiveness of these global initiatives in engaging local norms, institutions and practices remain contested.

The preceding chapters also show how sustainable seafood initiatives have negotiated global and local interests. First, market-based tools (Chapter 5) have been gradually amended to address local conditions better by adapting monitoring and evaluation practices to the actual conditions of producers, for example by hiring local auditors or modifying standards to allow for data-poor or group/community-based certification. Attempts to use ICT-based solutions for monitoring and evaluation have been put in place; these, while not necessarily adapting to local conditions, attempt to mitigate cost and weak producer capacity for compliance (e.g. Toonen and Bush, 2018). Second, the range of lifestyle and resource-based tools and arrangements outlined in Chapter 6 demonstrate alternative approaches to integrating 'local' norms into still globalised systems of governance, including fair, slow and community-supported seafood initiatives.

But to achieve the goals the movement has set out for itself, more work is needed on bridging the global/local divide. To do this, we argue, changes are required. First, greater attention is needed to engage with multiple nodes, made up of actors embedded within their own set of local dynamics and not just those of producers. This requires engaging with social relations that are often veiled at the global level – such as those relations that enable the persistence of illegal fishing or unfree labour in seafood value chains (Bailey et al., 2016b). Second, the movement needs to move beyond top-down and forensic assessments of sustainability. Instead, as outlined elsewhere (see Bush, 2016, 2017), new methodologies are needed that assess the extent to which influential actors, including buyers, NGOs and governments, *enable* others in the chain to improve their sustainability performance.

Public and private dynamics of sustainability governance

Setting and regulating ownership of seafood resources, be they related to fish stocks or the water, land, forest and biodiversity resources supporting aquaculture, are a central challenge for understanding the success of the sustainable seafood movement. The networked nature of the movement means that public and private actors continually collaborate to find solutions to the problems faced by both fisheries and aquaculture. In doing so, these actors experiment with the limits of their legitimacy for setting rules, the credibility of their sustainability claims and their authority for holding private and public actors to account.

The private initiatives established under the seafood movement have emerged out of a discontent over the inadequacy of state regulations in sustainably managing fishery and aquaculture resources. As seen in other food sectors (e.g. Boselie et al., 2003; Cashore et al., 2004, 2007; Pattberg, 2012; Abrams et al., 2018), the result is a range of interactions between the state and private seafood initiatives. In some cases private initiatives have supplemented government regulation, and sometimes put pressure on public authorities to fulfil their responsibilities through boycotts. The result is that the distinction between the roles of public and private actors in seafood governance is far less clear-cut than often assumed.

For example, states develop public standards to increase accessibility of local producers, to claim national sovereignty and to gain credibility in the market for existing government management (e.g. Foley, 2013). But states also use private standards to regain power over the private sector and other nations – as seen in the Pacific, where island states have used the MSC to 'recapture' ownership of tuna resources from distant-water nations and position themselves in the international tuna markets (Adolf et al., 2015). Complicating matters further, the strategies of NGOs and the private sector are also shifting. Certification has led to new spatial territories in oceans and coasts that cut through and across the national territories of states (Vandergeest et al., 2015). They also play a role in structuring negotiations in international RFMOs (Yeeting and Bush, 2019), and redefining the rules by which ecosystem protection, infrastructure and natural resource management are interrelated across landscapes (Bush et al., 2010; Ros-Tonen et al., 2015).

These intricate dynamics between public and private actors and regulation correspond to two ongoing debates.

First, it remains unclear whether innovative forms of public and private interaction effectively contribute to seafood sustainability. Multiple outcomes have been identified, ranging from a 'ratcheting up' of regulation and standards through a process of virtuous competition (Overdevest and Rickenbach, 2006; Bernstein and Cashore, 2007) to a race to the bottom if they compete on market share (Porter, 1999), or a stalemate if public or private initiatives block innovations in more effective regulatory approaches (Bitzer et al., 2008; Miller and Bush, 2015). The future of the sustainable seafood movement depends on serious engagement with the outcome of these different interactions. Moving forward, will growing influence of private actors undermine effective public policy, or will new modes of deliberation and partnership between public and private actors increase their impact?

Second, the role of public institutions in supporting the ongoing legitimacy of private tools and mechanisms needs more attention. The legitimacy of private initiatives remains precarious because they lack the formal procedures characteristic of government initiatives (Boström, 2006a, 2006b). These initiatives can increase legitimacy when they integrate existing public (national and local) policies and contribute expertise to technical areas of legalisation (Pattberg, 2006). However, such close collaboration comes at a risk when it is difficult to draw a clear boundary between public and private authority and responsibility (see e.g. Foley, 2013; Auld, 2014). The question is whether collaboration between public and private actors in the seafood movement will enhance or undermine the legitimacy of their shared initiatives.

Social dynamics of inclusion and exclusion

The tension between the inclusion and exclusion of producers against a measure of sustainability is at the heart of the so-called 'theory of change' of the sustainable seafood movement. As outlined by Roheim et al. (2018), this theory works on the basis of rewarding 'good' performers with entry to attractive markets and/or higher prices. Conversely, those producers that do not meet these requirements for sustainable production are excluded from markets with a demand for sustainably produced products. But despite its dominance in the seafood movement, this theory of change has been shown to be problematic for a number of reasons.

First, the ability of the seafood movement to be inclusive of a wide group of producers has undermined its overall credibility. As we have shown elsewhere (see Bush et al., 2013), the credibility of sustainability initiatives depends on justifiable decisions on who is included and who is excluded from (voluntarily) complying with sustainability requirements. If the sustainability initiatives of the seafood movement are to maintain their credibility, and as such their moral authority over the industry, they have to avoid a priori exclusion of any particular group of producers based on presumptions of impact or capability. If such presumptions are made, with the consequence of creating de facto barriers to trade, the impact of the movement as well as its overall moral authority may be undermined.

Second, inherent bias against some categories of fishers and fish farmers in public and private sustainability initiatives persists. As seen in a number of other sectors, market-oriented initiatives tend to have definitions of sustainability that are different to those of primary interest to producers, and are subject to 'market capture' rather than initiating equitable change (e.g. McCarthy, 2012; Brandi et al., 2015). But where such observations seem to hold for small-scale fishers (Bailey et al., 2016a; Crona et al., 2016a), they appear to be fraught in the context of aquaculture in developing countries, which is not easily categorised into 'industrial' and 'small scale'. Instead, the industry is increasingly driven by a broad category of medium-scale enterprises that do not easily fall into categories of small- or large-scale production (Belton et al., 2017). Thus not only is it important for the seafood movement to engage smallholders, but it must also better understand the increasingly ambiguous categorisation of producers in any future discussion on inclusiveness.

Third, the growing role of markets other than the EU and the US remains conspicuously underrepresented in the sustainable seafood movement. In spite of the considerable volumes entering international markets, the vast majority of seafood is consumed domestically – especially in developing countries (Belton et al., 2017). In the future this domestic consumption will likely rise further, driven by growing urbanisation. If current trends continue, it will also be supplemented by a more even trade between countries of the global South. The result will be a diversification of production and marketing into smaller, cheaper and relatively more accessible fish (e.g. Kassam and Dorward, 2017; Kaminski et al., 2018). While demand for sustainable products is yet to be exhibited in these markets, broader questions of sustainability that are yet to be taken up the seafood movement, such as the distribution of and access to the nutrients delivered by seafood, will be of ever greater importance in the future.

Taken together, these various insights call for a far more nuanced understanding of inclusion and exclusion, and ultimately the impact of the seafood movement, than the capabilities of producers to comply with standards alone.

Convergence versus divergence of sustainability tools and approaches

The tools and approaches used in the seafood movement have not only proliferated in number and type, but have also been replicated across different regions and countries. For example, recommendation lists are now used in over 15 countries,[1] 'better management practice' standards have been replicated by national and regional intergovernmental bodies, ecolabels have made small steps in Japan, Singapore and China, and product traceability is being rolled out in many parts of the world. Such replication appears to represent a gradual convergence of norms, practices and institutions and, as such, the extension of the sustainable seafood movement beyond the US and the EU.

But for all the potential of replicating these methods, an ongoing debate remains about the proliferation of the *types* of initiatives that make up the seafood movement. Proliferation can be seen as an evolutionary process of innovation (Overdevest and Zeitlin, 2014), or a function of a wider political economy of private foundations, NGOs and a search for competitive advantage in the retail sector (Konefal, 2013). Whatever the driver, it is unlikely that there will be fewer seafood initiatives in the near future, especially given the potential extension of the movement in emerging economies and the rise of public and industry initiatives as supplements or alternatives to NGO-led initiatives. As argued by Roheim et al. (2018), the ongoing proliferation of these initiatives can be positive provided some degree of alignment can be developed between them. At least two such forms of coordination appear feasible.

First, the greater involvement of states in harmonising sustainability initiatives could create pathways for improvement beyond single initiatives. Full harmonisation of private standards (or even national standards across regions) includes a commitment to comply with shared principles or norms and agreements on equivalence or mutual recognition. Where there is a regional intergovernmental body a degree of international harmonisation may even be possible, as already seen in the case of organic labels in the EU, for instance (Pekdemir, 2018; Zilberman et al., 2018). Alternatively, states could further pursue their own national certification initiatives or reaffirm the strength of their own fisheries' legislation and enforcement.

Second, market-based sustainable seafood aggregators (SSAs) could coordinate sustainability claims to reduce redundancy and increase effectiveness without constraining creative proliferation (Roheim et al., 2018). Unlike NGOs, these SSAs would not only steer purchasing decisions but also be exposed to the financial risk of poor advice. In practice this means that SSAs 'would provide recommendations to their clients on a portfolio of fisheries and aquaculture sources, with the quality of their portfolio varying much like a mutual fund with a portfolio of corporate stocks' (ibid.: 396). Supermarkets, processing companies and other providers of seafood would benefit from SSAs in two ways. First, SSAs may be more cost-effective than internalising the skills necessary to assess seafood initiatives. Second, they might increase the identification of producers of sustainable seafood and in doing so better represent the broader range of (small/large, developed/developing country) producers.

Some of the new initiatives outlined in the previous chapter have in fact started to fulfil some of these aggregator functions. For instance, the GSSI, GlobalGAP and some of the aquariums are taking on more direct economic risk for the advice they provide, based on their membership structures. But none has gone so far as to build the portfolios that could both provide access to a broad range of initiatives and, once involved, structure improvement across a range of 'tiered' producers and countries. The advantage of such an approach would be greater transparency about the sourcing of seafood, incentives for even poor sustainable seafood initiatives to include producers across the full spectrum of performance, the inclusion of a range of social and environmental norms (based on the portfolio's profile) and ultimately greater incentives to promote improvement towards sustainable forms of production.

Governing *through* value chains

All four dualisms reviewed above demonstrate the changing dynamics and emerging challenges of trying to steer an industry towards sustainability under conditions of globalisation. Many of the initiatives reviewed in the previous chapters have started to move away from 'demand drives improvement' theories of change to concentrate instead on affecting the social relations of production, trade and consumption beyond market transactions that shape the capabilities of seafood actors to improve. This is not to say that trade relations are not central to any analysis of the seafood movement. Instead, reflecting on the family of global commodity chain (GCC) and GVC/GPN theories (see Chapter 2), we argue it is necessary to understand the movement as made up of transactional actors, including producers, a variety of traders and consumers, and extra-transactional actors including government and NGOs. Doing so opens up a more sophisticated understanding of 'market-based' social movements that sees value chains as conduits for wider societal and environmental change.

The chains and networks literature (encompassed by the GCC, GVC and GPN frameworks) has made impressive gains in mapping out the spatial and organisational division of labour, the creation, capture and location of value, the effects of institutional and regulatory processes, and the influence of the wider cultural economy on economic organisation (Ponte and Sturgeon, 2013). Attention to sustainability has been limited to debate over how the economic order of production and consumption influences social equity and/or environmental performance. In most cases, however, sustainability has been dealt with as being eroded by capitalist modes of accumulation and neoliberal modes of control (e.g. Gibbon and Ponte, 2005; Neilson et al., 2014; Newsome et al., 2015; Alford and Phillips, 2018; Phelps et al., 2018). We argue that this neoliberal critique limits the possibility for new modes of governance that go beyond the market and enable firm and non-firm actors both internal and external to global chains to influence sustainable production.

Governance in the GCC/GVC/GPN literature has been variously defined in relation to a range of organisational, normative and regulatory processes. First, distinguishing between internal economic drivers within chains, and, in response to what Mayer and Gereffi (2010) label the 'governance deficit' of GVC research, the contribution of non-state and non-firm actors in defining the external rules through which chains are organised. Similarly, the GPN literature has only started to systematically address how, albeit without explicit reference to 'governance', (internal) firm strategy is influenced by a range of firm and non-firm interactions (Yeung and Coe, 2015). Building on earlier work (see Bush et al., 2015), we argue that this dichotomy of internal-external offers a useful starting point for understanding the emergence and interaction of sustainable seafood initiatives enacting 'market-based' modes of governance through global value chains.

Internal chain perspectives of governing sustainability can be divided into two types. First there is a functionalist type of (sustainable) 'supply chain' governance *within* chains (see Figure 7.1 for illustration). For instance, seafood processing companies have adopted internal governance initiatives as part of their wider logistics systems, including traceability for food safety and internal environmental water and energy-related management systems – such as compliance with the International Organization for Standardization (ISO) 14001 standard. The NGOs driving the sustainable seafood movement have tended not to focus on these internal forms of governance. The exclusion of such programmes is strategic, reflecting a division of responsibility between public and private sustainability issues. Debate has, however, continued on the consequences of not incorporating this internal dimension. As Pelletier and Tyedmers (2008) argued a decade ago, ecolabelling, certification and consumer education programmes remained incomplete because they did not incorporate criteria that represent the full life cycle of production, nor compared the performance between different forms of seafood. While academic studies incorporating such life-cycle criteria have proliferated in recent years (e.g. Avadí et al., 2018; Hilborn et al., 2018; Ziegler et al., 2018), they have been only partially taken up by the seafood movement – most notably in the creation of carbon footprint tools,[2] but not directly in ecocertification (Madin and Macreadie, 2015). Instead, the sustainable seafood movement has predominantly focused on the performance of producers in terms of their immediate environment – with the exception of aquaculture feed.

Second, the governance *of* chains refers to coordination mechanisms and power relations among economic actors, with particular attention given to the capability of suppliers to meet product specifications across global space. Recall from Chapter 2 that this organisation is most commonly characterised by five modes of GVC coordination by so-called lead firms (market, modular, relational, captive and hierarchy, based on Gereffi et al., 2005). By establishing this

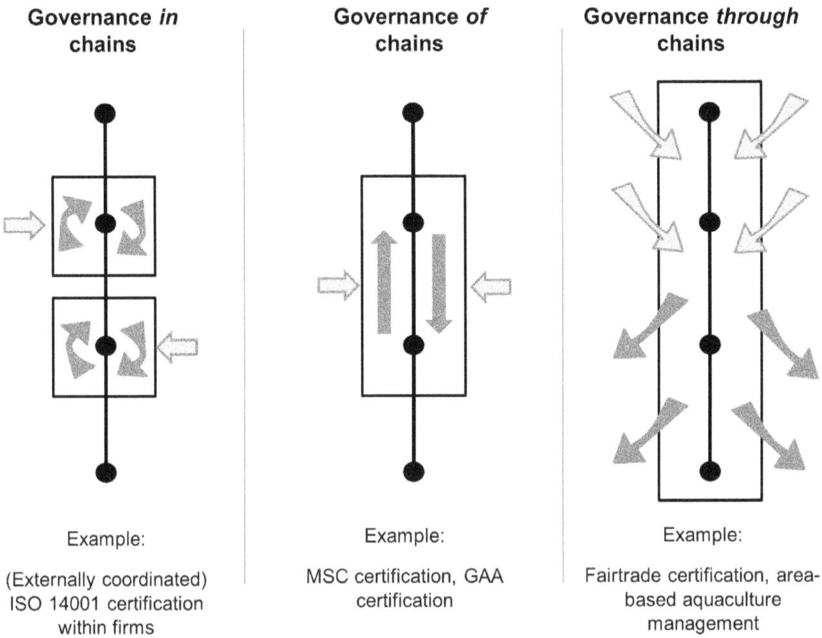

Figure 7.1 Illustration of three ideal types of governing chain sustainability.

Source: adapted from Bush et al. (2015).

Note

Light grey arrows entering chain indicate external governance related influences over firm actors. Dark grey arrows indicate firm-related governance influences internal and external to the chain.

control, lead firms influence the ability of buyers and suppliers to upgrade their production to meet product specifications, including credence attributes like sustainability. By codifying these attributes, private sustainability initiatives subject suppliers to product requirements that in turn structure the relationship buyers have with their suppliers. Yet, as explained throughout this book, the key critiques of these buyer codes (be they standards, recommendations or principles) are that they are focused on a narrow set of sustainability criteria related only to production; do not take up the local diversity of producers; focus on the delivery of sustainable products rather than sustainable production systems; and reinforce the power and control of buyers to the exclusion of (smallholder) producers.

Third, governing *through* chains recognises the 'external' influences on actors and processes from production to trade and consumption. As outlined in Chapter 2, these external influences include a range of horizontal (as opposed to vertical chain) relations with 'extra-transactional' actors – that is, NGOs and governments whose initiatives establish norms and standards for the performance of seafood production, trade and consumption. But governing through value chains also recognises the impact of sustainability outside the boundaries of direct economic activity to include a wider set of actors, environments and production processes beyond the simple performance of production. Such extended impact is increasingly evident in many of the 'beyond certification' initiatives outlined in Chapter 6. For instance, Fair Trade and the Slow Fish movement extend benefits to communities beyond fishers alone, and the various resource-based initiatives outlined aspire to ecosystem level impacts. Certification initiatives are also seeking to move beyond setting standards alone with programmes related to zonal management of aquaculture or engagement in regional improvement programmes (e.g. MSC's MedFish project[3]) which pre-assess and support the capacity of communities and governments to make sustainability improvements rather than directly supporting standard compliance.

Unlike the former two categories, governing through chains is understood as neither firm-level corporate social responsibility systems nor inter-firm coordination. Instead it is a broader level of governance that captures the interaction of the chain and its firm actors with a wider set of networked actors and activities that collectively steer sustainable production and consumption practices. It also invokes a systems approach that goes beyond industrial processes limited to site-specific labour or environmental performance by addressing the cumulative impact of sustainability throughout the chain. The seafood movement cannot then be thought of as a structural set of rules that prescribe action in the form of either internal environmental management or inter-firm coordination. It should instead be thought of as a set of production, trade, consumption and regulatory relations and practices that continually engage each other to shape sustainability outcomes. Such a perspective recasts the seafood chain as a global social phenomenon, made up of networked relations between market/non-market actors operating across the full spectrum of production, trade and consumption, and public/private actors engaging to steer towards improved forms of seafood production.

A networked 'governing through' understanding of social and economic relations holds three key consequences for how we understand market-based approaches to sustainability.

First, governing through value chains enables a systems perspective that goes beyond production only. One of the key constraints of internal approaches to value chain governance is the preoccupation with the performance of producers – as illustrated by the fact nearly all the initiatives reviewed in this book prescribe norms and principles for production. This productionist perspective is central to the logic of how lead firms establish control in value chains – paying lip service to societal norms such as sustainability but passing the burden of meeting these norms on to their suppliers, who in turn pass them on to producers. A key consequence, as already argued in this chapter, is an inherent bias towards producers that are able to comply to the exclusion of those that cannot, with the added consequence that sustainability will be as partial as the number of producers that can participate. By shifting attention to the conduct of other actors and their influence in and outside the chain, we can go beyond this productionist perspective to develop a (sea)food systems perspective to sustainability, which can subsequently expand our understanding of who influences sustainability outcomes beyond producers.

Second, a governing through value chains perspective highlights the multiple roles both chain and non-chain actors can play. If a chain is a conduit for social change, firms are not only profit seeking, NGOs not only driven by their moral authority and states not only (poorly performing) purveyors of regulation. Instead, firms, NGOs and states (and the individuals who enact them) have plural identities, norms and practices that lead them in designing, implementing and/or participating in sustainability initiatives. This plurality is exhibited by the emergence of pre-competitive partnerships and sustainability pledges and the uptake of social issues under the wide banner of seafood sustainability, while at the same fulfilling commercial roles in the market. It is also evidenced by the increasingly ambiguous role of NGOs as both standard setters and fee-based sustainability advisers, not unlike the SSAs outlined above, and by the role of states in taking a position in value chains through the application of private and public forms of ecocertification (Adolf et al., 2015; Foley and Havice, 2016). Identifying these multiple identities enables us to break predefined categories and assumptions about the role and function of actors in the sustainable seafood movement and helps us to recognise that these roles and functions have been and continue to be open to influence and change – to the benefit (or detriment) of sustainability.

Third, recognising governance through value chains gives attention to the networked relations between sustainability initiatives, be they either competitive or cooperative. The diverse values and norms on which the seafood movement is based (and has proliferated through) represent the imperfectly measurable nature of sustainability. Whether seen as a strategy by NGOs to capture market share or as representative of the diversity of locally embedded interpretations and values, the multiplicity of sustainability definitions creates competition. As illustrated above, such competition holds consequences for how public and private interact. But whatever the outcome, identifying the relations between initiatives enables us to identify how sustainability is shaped by a wider set of networked relations between 'extra-transactional' actors based on the exchange of knowledge, methodologies, norms and values in addition to economic interest. By identifying the nature of these relations, it is again possible to understand how they can be influenced for the betterment of the sustainable seafood movement as a whole.

These three consequences of a governing-through-value-chain perspective challenge us to develop a new theorisation of global food sustainability movements. Such a theory should go beyond production, recognise the multiple roles of actors and focus fundamentally on the interactions between actors and sustainability initiatives. It should also enable a more detailed understanding of what happens at any given node of the social networks that structure these interactions, and help us identify how sustainability initiatives recognise and shape the multiple flows of information, finance, regulation, norms, values and materials that influence the practice of sustainability.

Configuring seafood flows and practices

Recall from Chapter 2 that a flows and networks perspective is instructive for understanding the effectiveness of global sustainability movements for four reasons. First, it transcends old categories of state, market and civil society. Second, it does not limit societal interactions to geographically bounded systems or localities. Third, it enables us to understand sustainability outcomes as a function of social agency and interaction rather than as rules and regulation alone. And fourth, it allows us to extend the influence over sustainability outcomes to multiple, simultaneous social interactions that are both intentional and unintentionally influence 'better' practices far beyond the causal logics of market-based or value chain governance alone.

Also recall from Chapter 2 that a flows and networks perspective goes beyond conventional conceptualisations of sustainability as a function of technical processes and economic transactions. Instead, a flows perspective places attention on the sum of social interactions that constitute the social networks that interlink social movements and value chains. The flows that constitute these networked relations are both material (production inputs, energy, finances) and non-material (information about production and processing practices or environmental impact, governance arrangements and scientific and technological change) (Oosterveer, 2018). As Urry (2003) argues, these flows might be tightly coupled, multidimensional, enduring and predictable (such as state-based regulation and trade flows), or they may be transient and unpredictable (such as changes in food preferences, public opinion around sustainability or the financial trends of stock markets). Understanding how these flows, with their differing structure, predictability and persistence, interact and converge at different nodes and geographic locations enables us to understand more clearly not only who is included in or excluded from a network but also the composition of the relations which bring the members together or hold them apart.

Here we argue that a flows perspective enables a deeper understanding of how the composition of globally networked social relations influences the *practices* of social actors within any given node or geographic location. For instance, changes in the practices of producers, processors, traders or government employees to support the goal of sustainable seafood are determined by the confluence of different public and private rules, access to knowledge to use new technologies, changing global food preferences, public opinion and (restrictive) local economic interdependences. Identifying these multiple flows does not simply make things more complex; and importantly, and in contrast to Urry (ibid.), it also does not mean those subject to these flows are devoid of agency. Instead, as argued by Crossley (2018), it is precisely at the confluence of these multiple heterogeneous flows where social actors communicate and develop common understandings, rules and norms that determine if and how they act.

A flows perspective highlights how human agency is both held and shaped in common. This means that unlike perspectives focusing on individual choice and behavior (through for example economic incentives), flows highlight the multiple simultaneous networked social interactions that shape practices of production, trade or consumption (Spaargaren, 2011; Shove et al., 2012). These *social* practices are as such spatially and temporally dispersed doings and sayings that are organised by common understandings, goals and rules in a dialectic relationship with surrounding materials and (non-human) organisms (Schatzki, 1996). A social practice approach understands everyday doings and sayings as largely (unconsciously) routinised and socialised phenomena rather than reflecting conscious and rational decision-based individual behaviour. Individual behaviour is then the observable performance of these social practices, rather than a conscious expression of a single value, goal or relationship.

Understanding social practices as conditioned by flows and networks also challenges common functional categories and labels – such as production/producers or consumption/consumers. Instead, the social life of a practitioner (such as a producer) consists of a 'bundle' of social practices that go far beyond their functional (and reductionist) characterisation. For example, the performance of fish farming involves a series of interrelated practices directly related to fish production (e.g. stocking, harvesting, feeding) and trade (e.g. invoicing, ordering, reporting, packing, negotiating), as well as seemingly unrelated practices such as patronage, reciprocity, trust, information exchange and subsistence (Shove et al., 2012). If fish farming is reduced to a single practice (harvesting) or if that single practice is seen as isolated from other practices (e.g. negotiating, ordering, trust and reciprocity) which shape that practice, we are left with a partial understanding of what fish farming is in the first place (see Figure 7.2). We are also left with a limited range of flows (e.g. material, informational, financial, regulatory) that constitute and influence fish farming. In contrast, by understanding fish farming as constituted by a bundle of social practices, we can more accurately characterise what fish farming entails as well as the full range of influences (flows and practices) over how a material is used, a task undertaken or understanding expressed.

We argue that bringing flows and practices together enables a more precise understanding of how global social movements influence progress towards specific goals, such as seafood sustainability. When different social practices are linked, for example through networked seafood flows, they collectively configure the ways in which complex sustainable seafood provision is realised across global space. From this perspective the notion of linear supply chains is exploded to reveal a complex web of social practices within and external to nodes in the chain that are constituted by multiple global flows (see Figure 7.3). Each node is then more than an economic

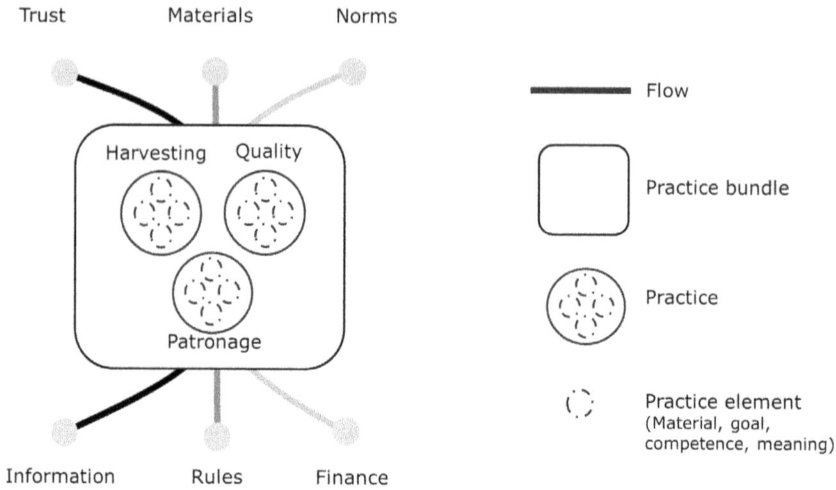

Figure 7.2 Composition of flows and practices.

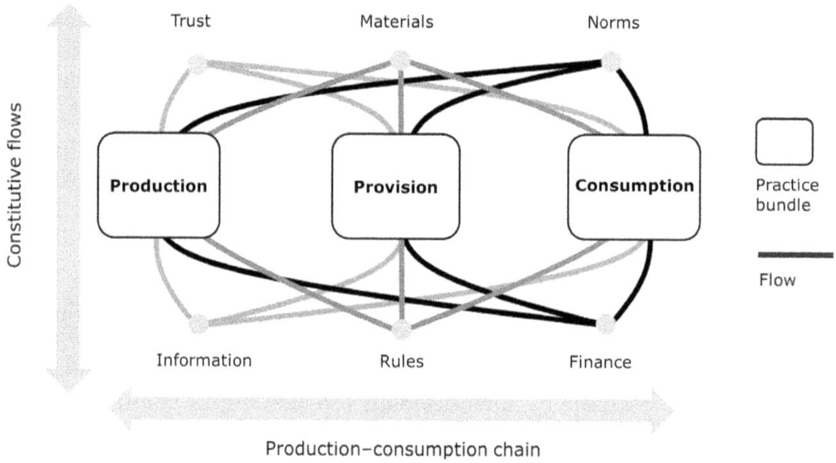

Figure 7.3 The interaction between global sustainability flows and practices.

transaction (shaped by product codification, economic risk and supplier capability; e.g. Gereffi et al., 2005). They are instead bundles of practices constituted by the meanings, materials, knowledge and goals set by and performed in response to flows of rules, finances, materials, energy, inputs, knowledge and moral expectations. The performance of these practices is also influenced by other practices in and around global supply chains that reconfigure the speed and content of global flows. For instance, practices related to provisioning can influence the flow of inputs, which in turn influence the knowledge required to meet the goals associated with the global sustainable seafood movement.

Based on this perspective, the governance of sustainable seafood is not determined by the sum of individual compliance to codes and standards. It is instead based on the degree of influence exercised over the flows that constitute any of the practices associated with the provision of sustainable seafood. Said differently, a shift to sustainable practices is determined by the ability of (would-be) governors to programme (in the words of Castells, 2009) the norms, values, incentives and materials that enable and/or disrupt (un)sustainable social practice. But it is also determined by the extent to which the internal organisation of practices in different sites and localities can be understood and 'orchestrated' to shape the movement of global flows so they enable the replication of sustainable practices.

By focusing on the interface of flows and practices we are forced to look at the totality of social processes that influence sustainability outcomes. A flows and practices perspective goes beyond rational or 'planned' behaviour, and instead understands that 'rationality' is socially mediated and born out of routine. Each observable outcome is a function of intended and unintended, conscious and unconscious social processes that influence the performance of sustainable or unsustainable practices. The perspective also shifts our attention away from governing individuals to governing the flows that influence action. In summary, by looking beyond the supply-drives-supply mode of the seafood movement, we can better understand what an unsustainable practice comprises, and in doing so refocus governance on to more influential flows that extend across global space.

In the rest of this chapter we go beyond these four commonly understood dimensions of global flows to illustrate how a flows and networks perspective combined with a (social) practices approach can enable a more dynamic understanding of how social sustainability movements are organised and operate over global space.

New governance configurations for sustainable seafood

So how would the flows and practices perspective outlined above offer an alternative understanding of the enactment of the seafood movement? And by understanding the seafood movement in more social terms, how might a new vision for governance be developed? The following subsections address these questions by 'reimagining' four governance configurations from a flows and practices perspective: state-based configurations, certification configurations, 'alternative' configurations and new territorial configurations. Each of these examples is chosen to illustrate the diverse ways in which flows and practices are mutually constitutive, and also to challenge our current understanding of the role and function of actors and global dynamics in the sustainable seafood movement.

State-based configurations

As illustrated throughout this book, the state plays a diverse set of roles in regulating global seafood. States regulating major importing markets like the US and the EU have overcome the national limits of their jurisdictions by regulating the performance of exporting states (and by default producers) by placing food safety and sustainability conditions on trade flows (e.g. IUU fishing, HACCP and traceability regulations). This places pressure on producer states to increase oversight and regulation of production. It also creates a fixation on the performance of producers which has driven much of the debate around the local/global, inclusion/exclusion dynamics outlined above. Moreover, this overemphasis on producers and production leads to an illusion of regulation – a regulation of the 'performance of performance', rather than the outcomes of production practices in their everyday context.

The flows and practices perspective moves beyond this state-based production/regulation perspective by drawing attention away from producer performance and towards the regulation of flows that constitute seafood production practices (see Figure 7.4).

Current fisheries' regulation focuses on regulating either input (such as fishing effort) or output (such as landings), both of which place fishers at the centre to be regulated as rational (economic) actors rather than socially embedded actors. A practices perspective starts with a focus on the fisher as the carrier of the practice, but also takes the practice as an entity constituted

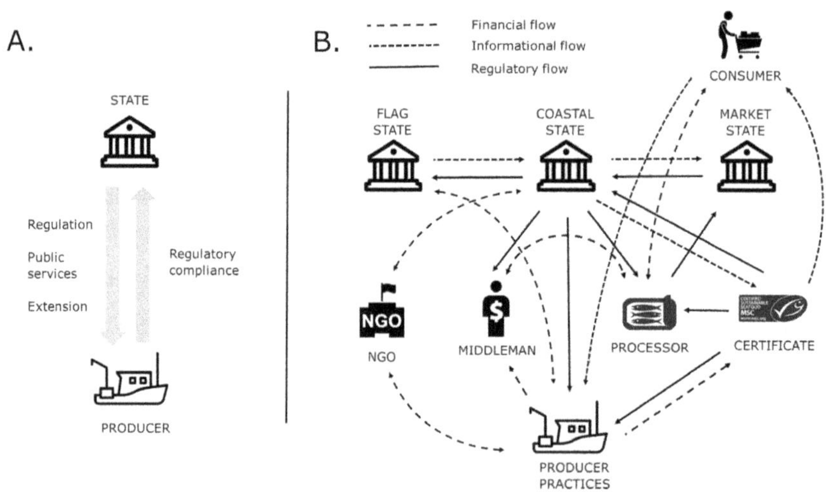

Figure 7.4 Reimagining state regulation from a flows and practices perspective.

Notes

A. Conventional state–producer relationship.
B. State-led configuration steering multiple regulatory and informational flows to shape the conditions under which trade and production are conducted.

by the confluence of multiple informational, regulatory, financial and material flows that are structured by an embedded set of social relations. To illustrate, fishing effort is performed by fishers, but is influenced by their level of indebtedness (Crona et al., 2016a); subsidies by different levels of government (Schuhbauer et al., 2017); the socialised routines of using fishing gear (Doddema et al., 2018); consumer (and therefore cultural) demands for some species over others; and competition (and conflict) between fishers over fishing grounds (Pomeroy et al., 2007). Perhaps an even better example is increasing focus on global financing rather than vessel activity in fighting IUU fishing. Galaz et al. (2018), for instance, estimated that 70 per cent of the known vessels implicated in IUU fishing are, or have been, flagged under a tax-haven jurisdiction. By advocating for states to mitigate the potential for tax havens, these authors are addressing a practice-shaping flow rather than IUU fishing as a performative outcome.

State-based aquaculture regulation faces similar limitations by focusing on the performance of producers rather than the flows that structure their practices. For example, the overapplication of pharmaceuticals by farmers is, at face value, in response to disease and can be regulated through a regime of post-harvest testing. But such 'output'-oriented regulation is only ever partial, in terms of both its coverage (particularly in production headed for domestic markets in developing countries) and addressing the rationality of farmers in applying these pharmaceuticals in the first place. A flows and practices perspective instead shifts attention to the indebtedness of farmers (Kusumawati et al., 2013), the trade relations that bind farmers to pharmaceutical companies and seed traders (Belton, 2012), social relations that manage water allocation (carrying disease) (Morgan et al., 2017), and the food and safety demands of export markets that reshape local access to resources (Mialhe et al., 2018).

Reimagining the state from a flows and practices perspective extends our understanding of regulation far beyond setting, enforcing and complying with regulations. This perspective reveals the current challenge faced by states to engage all the flows that simultaneously shape the collective practice of attaining sustainable seafood. But it also shows how states can more effectively engage with and influence different flows of finance, taxation, regulation and information to enable any shift to more sustainable practices.

Certification configurations

By providing information on the compliance of producers with a particular set of standards, certification schemes are thought to overcome information asymmetries around credence attributes such as sustainability, which then in turn allow for market demand to select environmentally sustainable products (Wessells et al., 2001). But from a flows and practices perspective such a linear understanding of markets and value chains veils a range of other pathways that certification creates for institutional change towards more environmentally sustainable modes of production. Beyond setting standards against which a producer's capacity for compliance can be assessed, certification initiatives purposefully coordinate many of the flows that can shape more sustainable seafood production practices.

Certification standard holders are therefore fundamentally networked organisations that not only link actors but also programme the confluence of information, regulation and financial and knowledge-based flows. In doing so they play a central governing role that is essential to their credibility and ultimately their effectiveness in fostering change towards sustainability beyond standard compliance alone (Gibbs, 2008; Eden, 2009). This networked organisational power is observed in both fisheries' and aquaculture certification initiatives in three ways.

First, the programming of global flows is increasingly evident in the day-to-day operations of certification initiatives. While standard holders should separate out standard development, compliance assessment and compliance development (as dictated by norms set by the ISO[4]), nearly all initiatives play a far more extended role than as just stagnant and distant owners of standards (as illustrated in Figure 7.5). Standard holders are increasingly engaging in research and development activities related to monitoring and evaluation, and they convene actors around fisheries to identify sustainability issues and (commonly) make them publicly available through their reporting system (Auld and Gulbrandsen, 2010). Indeed, the publicly available assessment reports (as well as public consultation submissions) by the MSC provide some of the most in-depth sources of information available on individual fisheries, bringing together multiple streams of data, scientific assessment and stakeholder

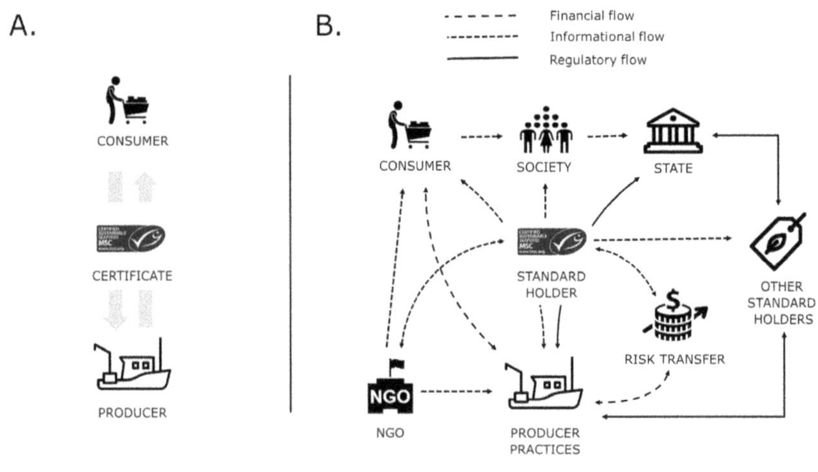

Figure 7.5 Reimagining certification from a flows and practices perspective.

Notes
A. Conventional certification–producer relation.
B. Networked configuration of certification.

experience. The reports of seafood recommendation initiatives, such as the Monterey Bay Aquarium, play a similar role for both fisheries and aquaculture (Kemmerly and Macfarlane, 2009). Seen as such, these initiatives are creating new levels of transparency around seafood production that, in principle, bring regulators and industry together to negotiate sustainability challenges.

Second, some certification organisations have explicitly extended their remit to shape advocacy and education for greater sustainability in the industry. For example, in the aquaculture sector the GAA has transformed its annual Global Outlook for Aquaculture Leadership conference and the publication of its magazine *The Aquaculture Advocate* to enable wider engagement with government and industry bodies around issues of sustainability that would not have been possible without setting standards for certification in the first place.

More radically, however, a flows and practices perspective allows us to re-vision the role of certification in organising and assessing more sustainable production practices, which challenges the dominant understanding of certification standards as a tool to set product specifications for producers in return for conditional market access. Many buyers, but also NGOs, states and industry associations, are in control of the means by which producers are able to comply with standards. These means can be understood as the confluence of a number of regulatory, licensing, informational and financial flows that enable or constrain producers to comply with standards. An alternative configuration for certification would be to refocus attention on the degree to which these actors programme such flows so that they create beneficial conditions for the performance of sustainable practices.

As outlined elsewhere (Bush, 2016, 2017), such a perspective can help us reverse the burden of proof for sustainability compliance, assessing and rewarding those actors with a certificate or ecolabel for promoting more effective modes of inclusive improvement. Not only would recognising and promoting continuous improvement towards sustainability increase the involvement of poorer-performing producers which cannot currently comply with certification, but it could also lead to a greater contribution to industry-wide sustainability because the cost of improving (and therefore the choice to do so) is no longer dispersed across literally millions of producers. In addition, shifting the cost of certification down the value chain would put pressure on retailers and suppliers to search for innovative and more efficient forms of organising seafood sustainability.

Alternative seafood configurations

The range of alternative seafood initiatives that have emerged throughout the seafood movement brings together a new set of actors, values and interests. As outlined above, these initiatives (extending from organic to Fair Trade, 'Slow Fish' and CSFs) have in many ways emerged in response to the local/global, inclusion/exclusion dynamics in global seafood provisioning. They also, in the words of Taylor (2005: 130), attempt to operate 'in the market' while not being 'of the market' – meaning they 'pursue alternative values and objectives such as social justice and environmental sustainability without being captured by the market's conventional logic, practices and dominant actors'. In doing so they bring together actors, logics and practices that seek to create wider environmental, cultural and community-level benefits through production practices – which in turn create 'alternative' understandings of sustainability beyond the environment alone.

A flows and practices perspective also provides an alternative configuration to the compliance logic of state and private forms of regulation. The 'alternative' tagline of these initiatives essentially indicates that knowledge, values and goals which are not usually associated with a capitalist mode of production and consumption are expressed by alternative product 'qualities', characterised by embeddedness, trust, place and the local (see for detail Goodman and Goodman, 2009). The configuration of these 'alternative' flows and networks can be formed in two distinct ways – both of which reflect an opposition to mainstream seafood markets.

First, these initiatives bring together flows of knowledge, expertise, social values and materials that shape the conditions through which alternative producer practices can be performed (see Figure 7.6). For instance, CSFs and the Slow Fish movement bring together local networks of producers and consumers in small and transparent markets. In doing so, they create new

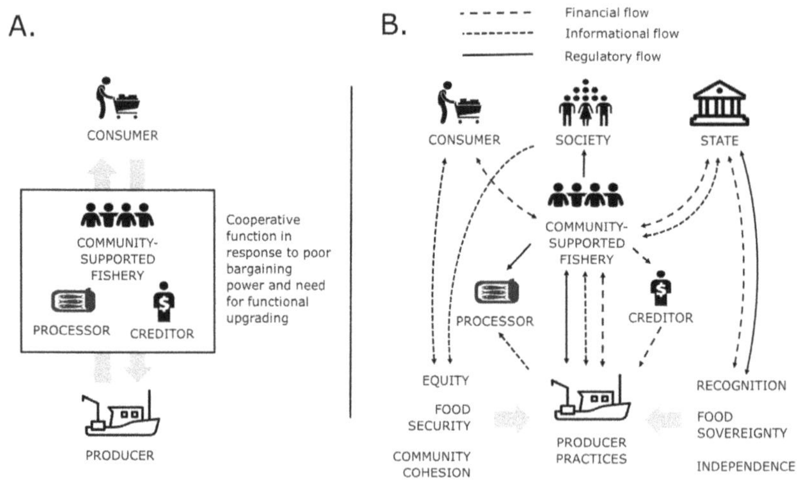

Figure 7.6 Reimagining alternative seafood networks from a flows and practices perspective.

Notes

A. Conventional market and regulatory relations marginalising small-scale fishers.

B. Community-supported fishery reorganises flows in support of alternative sustainability goals and meanings (values).

'meaning' around what 'local' or 'slow' signifies for producers, often including local regulators and industry associations, thereby enabling non-market outcomes related to recognition of work, independence and community building (Witter and Stoll, 2017), or even notions of social justice related to food security in regions predominantly geared to seafood exports (Loring et al., 2016). The construction of qualities remains centrally important for seafood, because it can be argued that freshness is a key consumer quality imbued with notions of 'local' (Campbell et al., 2013). But, more broadly, these initiatives also channel information, knowledge and markets for consumers to 'fight' mainstream supermarkets (Watts et al., 2018). Global initiatives, often operating through mainstream markets (e.g. organic and Fair Trade), similarly coordinate market and non-market actors, expertise, knowledge and materials that enable producers to engage in alternative production and consumers to engage through 'solidarity' movements underwritten by guarantees of social or nature-based outcomes. In all instances, however, these initiatives channel global and local flows to set the conditions for alternative modes of production, trade and consumption.

Second, similar to the 'governance' function of certification outlined above, these alternative initiatives act as boundary objects for wider societal debates that can lead to the transformation of mainstream modes of production and consumption. As argued by Whatmore et al. (2003: 390), initiatives steer alternative food networks by providing a coherent framing to stimulate their creation and internal structures that hold actors in conversation around a set of new production, trade and consumption practices 'without imposing a single order or design on what is being assembled' (also see Moschitz et al., 2015). Slow Fish is a case in point. Like the wider Slow Food movement of which it is a part, Slow Fish advocates taste, the importance of local culture and the empowerment of coastal fishing communities without prescribing the form or function of any one mode of production or trade. Instead Slow Fish creates a clear framing of alternative values and goals for fishers as a broadly defined community of producers, and in doing so starts a wider societal discussion around consumption relying on the persistence of these modes of production.

Reimagining these alternative sustainability initiatives through a flows and practices perspective reveals how new norms and values enable marginalised groups to shape opportunities effectively for maintaining lifestyles and improving market access. But they can also open up and engage with policy discussions around regionalism, identity and sovereignty – as seen in small-scale fisheries along the coast of the Mediterranean and in regions of North America such as Alaska and the maritime provinces of eastern Canada. It could also help better understanding of potential interaction and support for achieving alternative modes of seafood provisioning in collaboration with other sustainability initiatives – be it state regulation or private certification.

New territorial configurations

Finally, the role of initiatives in the seafood movement in coordinating diverse flows has led to the emergence of new territorial configurations located in and across state boundaries, within which the practices of seafood production are defined. In some instances these new territories are created by the definition of existing spatial units of production, such as aquaculture farms. As Vandergeest and Unno (2012) argue, aquaculture standards by definition set producer practices within the boundaries of farms, but in a way that has implications for control over inputs to these farms, such as water. But sustainability initiatives are now wielding their influence over new territorial configurations that extend beyond public space under national jurisdictions, as well as increasing control over how seafood production influences public space in the high seas. As the following two examples highlight, this process of territorialisation involves the coordination of flows that in turn shape sustainability practices.

First, sustainability initiatives such as certification are creating conditions for redefining regulatory territories in aquaculture by setting spatial boundaries around new values and goals in collaboration with states. For example, organic certification has led to state support for the expansion of global norms to regulate 'organic coasts' (Ha et al., 2012; Vandergeest et al., 2015). By bringing together NGOs, consultancies and processing companies, Naturland[5] certification was central in the creation of a complex series of nested spaces along the coast of Vietnam. Naturland standards sidestepped legal demands for mangrove cover and defined the farming unit through practice – that is, as 'a clearly marked managing sphere on which distinctly separate records are kept for inspection and documentation' (Naturland, 2018: 7). The government and NGOs alike have taken up this certification in the southern provinces of Vietnam to establish coastal landscape management that in many ways overrides provincial government regulations. Building on this and other cases (as elaborated in Chapter 6, and by Bottema et al., 2018), NGOs have taken up the development of such conceptualisations of area-based management to shape management of shared productions risks beyond the farm scale to achieve wider environmental goals – in some cases aspiring to use certification as a boundary object to configure these spatial territories.

Second, the case of digital enforcement using satellite- and drone-based monitoring and surveillance, outlined in Chapter 6, illustrates the creation of new regulatory territories in coastal and high-seas areas (see Figure 7.7). Recall that NGOs (including Global Fishing Watch and the Wildlife Conservation Society) have used new monitoring, control and surveillance technologies to collect spatially referenced data on the location of fishing vessels, gears and even fish stocks to eradicate IUU fishing activity (also see Toonen and Bush, 2018). These NGOs and technologies channel information between coastal and market states, as well as flag states that control the financing of IUU fishing vessels (Galaz et al., 2018). By essentially enclosing fishing

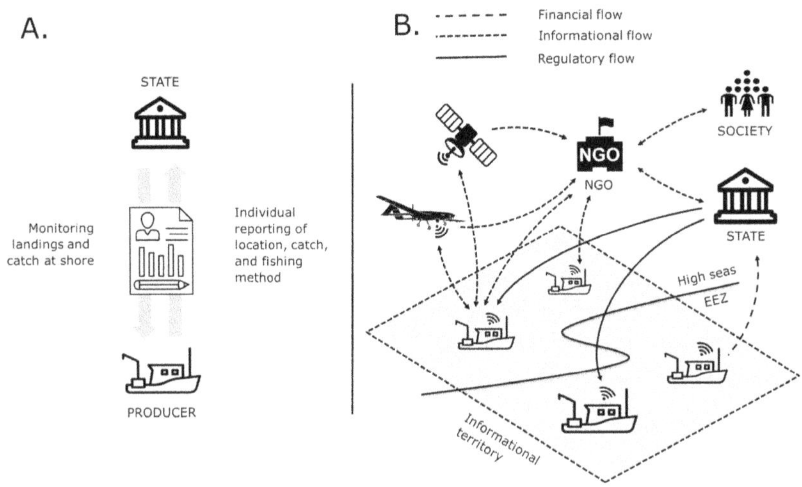

Figure 7.7 Reimagining digital enforcement territories from a flows and practices perspective.

Notes
A. Conventional state jurisdictions.
B. Flow-based boundary making.

grounds through surveillance, these NGOs and technologies are reconfiguring the roles and responsibilities of public and private actors involved in fisheries' management, including who collects and controls both legal and illegal fishing practice.

Conclusion

The central challenge of the sustainable seafood movement is to facilitate linkages in a broad global network of producers, traders and consumers through trade, advocacy and both state and private forms of regulation. Understanding the movement as constituted by a network made up of these multiple modes of engagement goes beyond market-based explanations of the seafood movement, and challenges any tendency to isolate individual initiatives as providing singular (either/or) solutions to seafood sustainability. We argue that understanding the seafood movement in these terms is limiting, because it reduces a range of fundamental social interactions down to a set of structural (rules) and transactions. These structural dimensions are reflected in the preoccupation of academic research with the ills of certification (which are in turn commonly reduced to the ills of neoliberal capitalism) and of practitioners with the building of even better assessment methodologies. A flows and practices perspective, in contrast, highlights how markets are made up of public and private social relations that structure global flows, and how these flows in turn shape the conditions in which sustainability practices are performed.

We also argue that a flows and practices perspective can change how sustainability initiatives operate. First, shifting attention to the flows that shape the (persistence of) existing practices instead of directly governing performance states can shape industry-wide change more effectively and efficiently. Second, any serious attempt to regulate flows instead of practices would require attention to a far wider range of actors than just producers. Third, the approach transcends the current notion of a public/private divide, which also focuses on the influence of private sustainability initiatives over the state. A flows perspective instead opens up the possibility that states engage in partnerships with any mix of private tools to enable them to reach their own goals both within their country and beyond. Finally, reflecting on the work of Sassen (2006), the perspective reorients our understanding of the state as a key actor that continues to shape the constitution of global flows, globalisation and the provision of sustainable seafood.

Notes

1 These recommendation lists are in most cases based on the US Monterey Bay Aquarium Seafood Watch and Dutch Good Fish Guide methodologies.

2 See for example the carbon footprint tool developed by the UK industry body Seafish, available at www.seafish.org/GHGEmissionsProfiler/v1/ (accessed 17 November 2018).

3 More detail on the MedFish project is available at www.project-medfish.com/ (accessed 21 November 2018).

4 The ISO has published a series of standards starting from the 1990s on the conduct of standard holders and conformity assessment organisations. These standards have focused on the expected competencies of both kinds of organisations, as well as how they should organise the avoidance of (commercial) conflicts of interest between their respective roles. A catalogue of these standards is available at www.iso.org/ics/03.120.20/x/ (accessed 14 August 2018).

5 Naturland is a German association certifying organic producers around the world. It has developed a standard for organic aquaculture that is applied in several countries in Southeast Asia. See www.naturland.de/en/naturland/naturland-standards/285-naturland-uk/naturland/naturland-standards/standards-producers/851-aquaculture.html (accessed 17 November 2018).

References

Abrams, J., E. Nielsen, D. Diaz, T. Selfa, E. Adams, J. L. Dunn and C. Moseley (2018). 'How do states benefit from nonstate governance? Evidence from forest sustainability certification', *Global Environmental Politics* 18(3), 66–85.

Adolf, S., S. R. Bush and S. Vellema (2015). 'Reinserting state agency in global value chains: The case of MSC certified skipjack tuna', *Fisheries Research* 182, 79–87.

Alford, M. and N. Phillips (2018). 'The political economy of state governance in global production networks: Change, crisis and contestation in the South African fruit sector', *Review of International Political Economy* 25(1), 98–121.

Auld, G. (2014). *Constructing Private Governance: The Rise and Evolution of Forest, Coffee, and Fisheries Certification*. New Haven, CT: Yale University Press.

Auld, G. and L. H. Gulbrandsen (2010). 'Transparency in nonstate certification: Consequences for accountability and legitimacy', *Global Environmental Politics* 10(3), 97–119.

Avadí, A., P. J. Henriksson, I. Vázquez-Rowe and F. Ziegler (2018). 'Towards improved practices in life cycle assessment of seafood and other aquatic products', *International Journal of Life Cycle Assessment* 23(5), 979–981.

Bailey, M., S. R. Bush, P. Oosterveer and L. Larastiti (2016a). 'Fishers, Fair Trade, and finding middle ground', *Fisheries Research* 182(Supplement C), 59–68.

Bailey, M., S. R. Bush, A. Miller and M. Kochen (2016b). 'The role of traceability in transforming seafood governance in the global South', *Current Opinion in Environmental Sustainability* 18, 25–32.

Belton, B. (2012). 'Culture, social relations and private sector development in the Thai and Vietnamese fish hatchery sectors', *Asia Pacific Viewpoint* 53(2), 133–146.

Belton, B., S. R. Bush and D. Little (2017). 'Not just for the wealthy: Rethinking farmed fish consumption in the global South', *Global Food Security* 16, 85–92.

Bernstein, S. and B. Cashore (2007). 'Can non-state global governance be legitimate? An analytical framework', *Regulation and Governance* 1, 1–25.

Bitzer, V., M. Francken and P. Glasbergen (2008). 'Intersectoral partnerships for a sustainable coffee chain: Really addressing sustainability or just picking (coffee) cherries?', *Global Environmental Change* 18(2), 271–284.

Boselie, D., S. Henson and D. Weatherspoon (2003). 'Supermarket procurement practices in developing countries: Redefining the roles of the public and private sectors', *American Journal of Agricultural Economics* 85(5), 1155–1161.

Boström, M. (2006a). 'Establishing credibility: Practising standard-setting ideals in a Swedish seafood-labelling case', *Journal of Environmental Policy & Planning* 8(2), 135–158.

Boström, M. (2006b). 'Regulatory credibility and authority through inclusiveness: Standardization organizations in cases of eco-labelling', *Organization* 13(3), 345–367.

Bottema, M., S. Bush and P. Oosterveer (2018). 'Moving beyond the shrimp farm: Spaces of shared environmental risk?', *Geographical Journal* online, DOI https://doi.org/10.1111/geoj.12280.

Brandi, C., T. Cabani, C. Hosang, S. Schirmbeck, L. Westermann and H. Wiese (2015). 'Sustainability standards for palm oil: Challenges for smallholder certification under the RSPO', *Journal of Environment & Development* 24(3), 292–314.

Bush, S. R. (2016). 'Reversing the burden of proof for sustainable aquaculture', *Solutions* 7(4), 27–30.

Bush, S. R. (2017). 'Certify sustainable retailers?', in J. Duncan and M. Bailey (eds) *Sustainable Food Futures: Multidisciplinary Solutions*. London: Routledge, 133–144.

Bush, S., H. Toonen, P. Oosterveer and A. Mol (2013). 'The "devil's triangle" of MSC certification: Balancing credibility, accessibility and continuous improvement', *Marine Policy* 37, 288–293.

Bush, S. R., P. Oosterveer, M. Bailey and A. P. Mol (2015). 'Sustainability governance of chains and networks: A review and future outlook', *Journal of Cleaner Production* 107, 8–19.

Bush, S. R., P. A. van Zwieten, L. Visser, H. van Dijk, R. Bosma, W. F. de Boer and M. Verdegem (2010). 'Scenarios for resilient shrimp aquaculture in tropical coastal areas', *Ecology and Society* 15(2), 15.

Campbell, L. M., N. Boucquey, J. Stoll, H. Coppola and M. D. Smith (2013). 'From vegetable box to seafood cooler: Applying the community-supported agriculture model to fisheries', *Society & Natural Resources* 27(1), 88–106.

Cashore, B., G. Auld and D. Newsom (2004). *Governing Through Markets: Forest Certification and the Emergence of Non-State Authority*. New Haven, CT: Yale University Press.

Cashore, B., G. Auld, S. Bernstein and C. McDermott (2007). 'Can non-state governance "ratchet up" global environmental standards? Lessons from the forest sector', *Review of European Community & International Environmental Law* 16(2), 158–172.

Castells, M. (2009). *Communication Power*. Oxford: Oxford University Press.

Crona, B. I., X. Basurto, D. Squires, S. Gelcich, T. M. Daw, A. Khan, E. Havice, V. Chomo, M. Troell and E. A. Buchary (2016a). 'Towards a typology of interactions between small-scale fisheries and global seafood trade', *Marine Policy* 65, 1–10.

Crona, B. I., T. M. Daw, W. Swartz, A. V. Norström, M. Nyström, M. Thyresson, C. Folke, J. Hentati-Sundberg, H. Österblom and L. Deutsch (2016b). 'Masked, diluted and drowned out: How global seafood trade weakens signals from marine ecosystems', *Fish and Fisheries* 17(4), 1175–1182.

Crossley, N. (2018). 'Networks, interactions and relations', in F. Dépelteau (ed.) *The Palgrave Handbook of Relational Sociology*. Cham: Springer International, 481–498.

Doddema, M., G. Spaargaren, B. Wiryawan and S. R. Bush (2018). 'Fisher responses to private monitoring interventions in an Indonesian tuna handline fishery', *Fisheries Research* 208, 49–57.

Dolan, C. S. (2010). 'Virtual moralities: The mainstreaming of Fairtrade in Kenyan tea fields', *Geoforum* 41(1), 33–43.

Eden, S. (2009). 'The work of environmental governance networks: Traceability, credibility and certification by the Forest Stewardship Council', *Geoforum* 40(3), 383–394.

Foley, P. (2013). 'National government responses to Marine Stewardship Council (MSC) fisheries certification: Insights from Atlantic Canada', *New Political Economy* 18(2), 284–307.

Foley, P. and E. Havice (2016). 'The rise of territorial eco-certifications: New politics of transnational sustainability governance in the fishery sector', *Geoforum* 69, 24–33.

Galaz, V., B. Crona, A. Dauriach, J.-B. Jouffray, H. Österblom and J. Fichtner (2018). 'Tax havens and global environmental degradation', *Nature Ecology & Evolution* 2(9), 1352–1357.

Gereffi, G., J. Humphrey and T. Sturgeon (2005). 'The governance of global value chains', *Review of International Political Economy* 12(1), 78–104.

Gibbon, P. and S. Ponte (2005). *Trading Down. Africa, Value Chains, and the Global Economy*. Philadelphia, PA: Temple University Press.

Gibbs, M. T. (2008). 'Network governance in fisheries', *Marine Policy* 32(1), 113–119.

Goodman, D. and M. Goodman (2009). 'Alternative food networks', *International Encyclopedia of Human Geography* 3, 208–220.

Gudmundsson, E. and C. R. Wessells (2000). 'Ecolabeling seafood for sustainable production: Implications for fisheries management', *Marine Resource Economics* 15(2), 97–113.

Ha, T. T. T., S. R. Bush, A. P. Mol and H. van Dijk (2012). 'Organic coasts? Regulatory challenges of certifying integrated shrimp–mangrove production systems in Vietnam', *Journal of Rural Studies* 28(4), 631–639.

Hilborn, R., J. Banobi, S. J. Hall, T. Pucylowski and T. E. Walsworth (2018). 'The environmental cost of animal source foods', *Frontiers in Ecology and the Environment* 16(6), 329–335.

Kaminski, A. M., S. Genschick, A. S. Kefi and F. Kruijssen (2018). 'Commercialization and upgrading in the aquaculture value chain in Zambia', *Aquaculture* 493, 355–364.

Kassam, L. and A. Dorward (2017). 'A comparative assessment of the poverty impacts of pond and cage aquaculture in Ghana', *Aquaculture* 470, 110–122.

Kemmerly, J. D. and V. Macfarlane (2009). 'The elements of a consumer-based initiative in contributing to positive environmental change: Monterey Bay Aquarium's Seafood Watch program', *Zoo Biology* 28(5), 398–411.

Konefal, J. (2013). 'Environmental movements, market-based approaches, and neoliberalization: A case study of the sustainable seafood movement', *Organization & Environment* 26(3), 336–352.

Kusumawati, R., S. R. Bush and L. E. Visser (2013). 'Can patrons be bypassed? Frictions between local and global regulatory networks over shrimp aquaculture in east Kalimantan', *Society & Natural Resources* 26(8), 898–911.

Loring, P. A., S. C. Gerlach and H. L. Harrison (2016). 'Seafood as local food: Food security and locally caught seafood on Alaska's Kenai Peninsula', *Journal of Agriculture, Food Systems, and Community Development* 3(3), 13–30.

Madin, E. M. P. and P. I. Macreadie (2015). 'Incorporating carbon footprints into seafood sustainability certification and eco-labels', *Marine Policy* 57, 178–181.

Mayer, F. and G. Gereffi (2010). 'Regulation and economic globalization: Prospects and limits of private governance', *Business and Politics* 12(3), 1–25.

McCarthy, J. F. (2012). 'Certifying in contested spaces: Private regulation in Indonesian forestry and palm oil', *Third World Quarterly* 33(10), 1871–1888.

Mialhe, F., E. Morales, S. Dubuisson-Quellier, I. Vagneron, L. Dabbadie and D. Little (2018). 'Global standardization and local complexity. A case study of an aquaculture system in Pampanga Delta, Philippines', *Aquaculture* 493, 365–375.

Miller, A. M. and S. R. Bush (2015). 'Authority without credibility? Competition and conflict between ecolabels in tuna fisheries', *Journal of Cleaner Production* 107, 137–145.

Morgan, M., G. Terry, S. Rajaratnam and J. Pant (2017). 'Socio-cultural dynamics shaping the potential of aquaculture to deliver development outcomes', *Reviews in Aquaculture* 9(4), 317–325.

Moschitz, H., D. Roep, G. Brunori and T. Tisenkopfs (2015). 'Learning and innovation networks for sustainable agriculture: processes of co-evolution, joint reflection and facilitation', *Journal of Agricultural Education and Extension* 21(1), 1–11.

Mutersbaugh, T. (2005). 'Just-in-space: Certified rural products, labor of quality, and regulatory spaces', *Journal of Rural Studies* 21(4), 389–402.

Naturland (2018) 'Naturland standards organic aquaculture', Version 06/18. Gräfelfing: Naturland.

Neilson, J. (2014). 'Value chains, neoliberalism and development practice: The Indonesian experience', *Review of International Political Economy* 21(1), 38–69.

Neilson, J., B. Pritchard and H. W.-C. Yeung (2014). 'Global value chains and global production networks in the changing international political economy: An introduction', *Review of International Political Economy* 21(1), 1–8.

Newsome, K., P. Taylor, J. Bair and A. Rainnie (2015). *Putting Labour in Its Place: Labour Process Analysis and Global Value Chains*. London and New York: Palgrave.

Oosterveer, P. (2018). 'Global environmental networks and flows addressing global environmental change', in M. Bostrom and D. J. Davidson (eds) *Environment and Society*. London: Palgrave Macmillan, 95–118.

Overdevest, C. and M. G. Rickenbach (2006). 'Forest certification and institutional governance: An empirical study of Forest Stewardship Council certificate holders in the United States', *Forest Policy and Economics* 9(1), 93–102.

Overdevest, C. and J. Zeitlin (2014). 'Assembling an experimentalist regime: Transnational governance interactions in the forest sector', *Regulation & Governance* 8(1), 22–48.

Pattberg, P. (2006). 'Private governance and the South: Lessons from global forest politics', *Third World Quarterly* 27(4), 579–593.

Pattberg, P. H. (2012). *Public–Private Partnerships for Sustainable Development: Emergence, Influence and Legitimacy*. Cheltenham: Edward Elgar Publishing.

Pekdemir, C. (2018). 'On the regulatory potential of regional organic standards: Towards harmonization, equivalence, and trade?', *Global Environmental Change* 50, 289–302.

Pelletier, N. and P. Tyedmers (2008). 'Life cycle considerations for improving sustainability assessments in seafood awareness campaigns', *Environmental Management* 42(5), 918–931.

Phelps, N. A., M. Atienza and M. Arias (2018). 'An invitation to the dark side of economic geography', *Environment and Planning A* 50(1), 236–244.

Platteau, J. P. (1994). 'Behind the market stage where real societies exist – Part I: The role of public and private order institutions', *Journal of Development Studies* 30(3): 533–577.

Pomeroy, R., J. Parks, R. Pollnac, T. Campson, E. Genio, C. Marlessy, E. Holle, M. Pido, A. Nissapa and S. Boromthanarat (2007). 'Fish wars: Conflict and collaboration in fisheries management in Southeast Asia', *Marine Policy* 31(6), 645–656.

Ponte, S. (2002). 'Brewing a bitter cup? Deregulation, quality and the re-organization of coffee marketing in East Africa', *Journal of Agrarian Change* 2(2), 248–272.

Ponte, S. and T. Sturgeon (2013). 'Explaining governance in global value chains: A modular theory-building effort', *Review of International Political Economy* 21, 1–29.

Porter, G. (1999). 'Trade competition and pollution standards: "Race to the bottom" or "stuck at the bottom"?', *Journal of Environment & Development* 8(2), 133–151.

Roheim, C. A., S. R. Bush, F. Asche, J. N. Sanchirico and H. Uchida (2018). 'Evolution and future of the sustainable seafood market', *Nature Sustainability* 1(8), 392–398.

Ros-Tonen, M. A., Y.-P. B. Van Leynseele, A. Laven and T. Sunderland (2015). 'Landscapes of social inclusion: Inclusive value-chain collaboration through the lenses of food sovereignty and landscape governance', *European Journal of Development Research* 27(4), 523–540.

Sassen, S. (2006). *Territory, Authority, Rights: From Medieval to Global Assemblages*. Princeton, NJ and Oxford: Princeton University Press.

Schatzki, T. R. (1996). *Social Practices: A Wittgensteinian Approach to Human Activity and the Social*. Cambridge: Cambridge University Press.

Schuhbauer, A., R. Chuenpagdee, W. W. L. Cheung, K. Greer and U. R. Sumaila (2017). 'How subsidies affect the economic viability of small-scale fisheries', *Marine Policy* 82, 114–121.

Shove, E., M. Pantzar and M. Watson (2012). *The Dynamics of Social Practice: Everyday Life and How It Changes*. London: Sage.

Spaargaren, G. (2011). 'Theories of practices: Agency, technology, and culture. Exploring the relevance of practice theories for the governance of sustainable consumption practices in the new world-order', *Global Environmental Change* 21(3), 813–822.

Taylor, P. L. (2005). 'In the market but not of it: Fair Trade coffee and Forest Stewardship Council certification as market-based social change', *World Development* 33(1), 129–147.

Taylor, P. L., D. L. Murray and L. T. Raynolds (2005). 'Keeping trade fair: Governance challenges in the Fair Trade coffee initiative', *Sustainable Development* 13, 109–208.

Toonen, H. M. and S. R. Bush (2018). 'The digital frontiers of fisheries governance: Fish attraction devices, drones and satellites', *Journal of Environmental Policy & Planning* online, DOI https://doi.org/10.1080/1523908X.2018.1461084.

Tsing, A. L. (2011). *Friction: An Ethnography of Global Connection*. Princeton, NJ: Princeton University Press.

Urry, J. (2003). *Global Complexity*. Cambridge: Polity Press.

Vandergeest, P., S. Ponte and S. Bush (2015). 'Assembling sustainable territories: Space, subjects, objects, and expertise in seafood certification', *Environment and Planning A* 47(9), 1907–1925.

Vandergeest, P. and A. Unno (2012). 'A new extraterritoriality? Aquaculture certification, sovereignty, and empire', *Political Geography* 31(6), 358–367.

Watts, D., J. Little and B. Ilbery (2018). '"I am pleased to shop somewhere that is fighting the supermarkets a little bit." A cultural political economy of alternative food networks', *Geoforum* 91, 21–29.

Wessells, C., K. Cochrane, C. Deere, P. Wallis and R. Willmann (2001). *Product Certification and Ecolabelling for Fisheries Sustainability*. Rome: FAO.

Whatmore, S., P. Stassart and H. Renting (2003). *What's Alternative about Alternative Food Networks?* London: Sage.

Witter, A. and J. Stoll (2017). 'Participation and resistance: Alternative seafood marketing in a neoliberal era', *Marine Policy* 80, 130–140.

Yeeting, A. and S. R. Bush (2019). 'RFMO-MSC smart regulatory mixes for transboundary tuna fisheries', in J. Van Erp, M. Faure, A. Nollkaemper and P. Niels (eds) *Smart Mixes in Relation to Transboundary Environmental Harm*. Cambridge: Cambridge University Press.

Yeung, H. W.-C. and N. M. Coe (2015). 'Toward a dynamic theory of global production networks', *Economic Geography* 91, 29–58.

Ziegler, F., E. A. Groen, S. Hornborg, E. A. Bokkers, K. M. Karlsen and I. J. de Boer (2018). 'Assessing broad life cycle impacts of daily onboard decision-making, annual strategic planning, and fisheries management in a northeast Atlantic trawl fishery', *International Journal of Life Cycle Assessment* 23(7), 1357–1367.

Zilberman, D., S. Kaplan and B. Gordon (2018). 'The political economy of labeling', *Food Policy* 78, 6–13.

8 The future of sustainable seafood

Introduction

In the preceding chapters we described the global dynamics that underpin global seafood production and consumption. We also reviewed the wide variety of initiatives taken by societal actors to advance sustainability in global seafood supply. In doing so, we demonstrated the enduring importance of the state in steering towards sustainable production, the diversity of sustainability claims made by NGOs and the range of 'theories of change' employed by all actors involved under the banner of the sustainable seafood movement. Based on the insights collected from the previous chapters, we argue for a reconceptualisation of the movement in terms of flows and practices that can think beyond economic and technical dimensions of sustainability and instead identify the social dynamics that enable or constrain change throughout the global seafood system.

In this concluding chapter we synthesise the overarching insights gained from our journey through the sustainable seafood movement. We first summarise the challenges for future sustainable seafood provisioning, and then outline what we see as the overarching themes for further research on how the seafood movement can be more effective in transitioning the 'seafood system' towards sustainability. We argue that the governance of this seafood system requires the incorporation of social relations (and their intersecting flows) of production, trade and consumption that ultimately determine sustainable seafood practices. By doing so, the movement can evolve into a more encompassing and integrated set of governance arrangements involving diverse sets of state, market and civil society actors operating in and across multiple scales and localities. We conclude by formulating seven propositions for the further development of the sustainable seafood movement.

Challenges for future sustainable seafood provision

The previous chapters covered the multiple challenges faced by the sustainable seafood movement in fulfilling its mission of steering producers and governments to improve their stewardship of coastal and marine environments and communities. Here we synthesise these into four of the most important insights that will affect the effectiveness of the movement going forward.

First, the OECD markets currently demanding sustainable seafood will not be the main sites of seafood consumption in the future. This is already the case for aquaculture, where nine out of ten fish produced are consumed domestically in the global South. But a similar trend is likely for international trade, as the rising middle-class populations in Asia increase their demand for fish and other aquatic organisms (Béné et al., 2015; Kobayashi et al., 2015). For the sustainable seafood movement to remain relevant, far more consideration needs to be given to the potential for sustainable consumerism within Asia (see for example Fabinyi and Liu, 2014; Fabinyi, 2016), as well as the (emerging) seafood markets of Africa and Latin America. Examples of such engagement are already available: the MSC has entered the Japanese and Chinese markets, the Monterey Bay Seafood Watch programme has staff based throughout Asia and Latin America, and the annual Tokyo Sustainable Seafood Symposium has been held in Japan since 2015.[1] But to be successful going forward, the movement will need to expand its geographical reach significantly to include the global South in setting the future sustainability agenda.

Second, there is a need for better integration between public and private sustainability initiatives. As outlined throughout this book, the range of state, market and civil-society-led initiatives that group themselves under the banner of the sustainable seafood movement contribute a rich mosaic of approaches that, if effectively coordinated (e.g. cross-referenced, harmonised or integrated), can reinforce each other (Gulbrandsen, 2014). Determining how such alignment or reinforcement can occur has received far less attention in the seafood movement than in other sectors, such as forestry and agriculture (e.g. Pattberg, 2007; Overdevest and Zeitlin, 2014). Various proposals and attempts to improve coordination are available, including the development of SSAs and greater state oversight and assurance mechanisms (e.g. Roheim et al., 2018). But whether and how they can be effective are an ongoing area of research.

Third, transparency and accountability remain central to the ongoing success of the sustainable seafood movement as a whole. As argued by Pramod et al. (2014: 111),

> the gaps in the system occur at many levels: at sea, where monitoring, control and surveillance frequently remain inadequate; in ports, where systems to document catch landings are often weak or non-transparent; and in market countries, where effective systems to require traceability and proof of legal origin are lacking.

But as demands to close these gaps grow, the ability of NGO and state sustainability initiatives to deliver has also increased. As outlined in various chapters of this book, value chain and remote-sensing technologies applied to seafood have started to make promises of what can be termed 'extreme disclosure' (e.g Lewis and Boyle, 2017; Toonen and Bush, 2018). But any NGO applying these technologies will have to ensure that they lead to changes in production practices of a broad spectrum of (large/small) producers to remain credible.

Fourth, there is a need to expand international cooperation for sustainability beyond the high seas. International cooperation for fisheries focuses on highlight migratory fish stocks through the UN Law of the Sea and Transboundary Fish Stocks Agreement. RFMOs may be mandated to assess stocks and set limits and reference target points for the management of stocks. But their focus is on transboundary and high-seas fish stocks that make up only 4.2 per cent of annual marine-capture fisheries' production and 2.4 per cent of total seafood production (Schiller et al., 2018). The rest is caught within national jurisdictions. Aquaculture, despite its considerable contribution to animal protein provision into the future, has no international platform for cooperation. This limited scope of international cooperation beyond the high seas does not reflect the full impact of the global seafood sector, including, for example, the global scope of aquaculture feed and fish-processing waste flows and the global interdependence on fish-food protein. Reimagining how states could cooperate through some kind of 'international panel on sustainable seafood' (in whatever form) may help in driving a global sustainability agenda forward.

Finally, a more coherent balance is required between 'seafood as food' and 'seafood for ocean wealth'.[2] Key questions are already being asked about whether the environmental goals of the seafood movement have placed too much emphasis on improving ocean health through the narrow lens of fish stocks and fish farming. Understanding seafood as part of a wider global food system may instead provide a more integrative approach for linking a broad set of sustainability issues to seafood consumption. Not only would a food systems' approach bring into sharp focus the role of seafood as an important source of nutrition around the world, but it would also better link the practice of seafood consumption to, for instance, biodiversity impacts, labour and carbon emissions in both the world's oceans and terrestrial agriculture systems. Central to such an approach, and highly complementary to the flows and practices approach introduced in the previous chapter, are links to aquaculture feed ingredients and the impact of dietary shifts between different forms of protein. Seafood sustainability would then be placed in the broader context of food systems in general, and the sustainability of the industry would be related to the global challenges of sustainably feeding 9 billion people by 2050.

Building integrated sustainable seafood governance arrangements

Understanding the sustainable seafood movement as a combination of socially networked flows and practices can help to overcome many of the challenges outlined above. By breaking the sustainable seafood movement into a constituent set of social and material flows, and by understanding how these flows structure and respond to production, trade and consumption practices, we offer a new way of understanding the form and function of global sustainability movements. In developing the flows and practices approach we also hope to stimulate debate on the future design of the seafood movement and the sustainability initiatives it has fostered. Here we identify seven propositions that can shape debate over the future design of the movement.

First, *understanding the seafood movement as the composition of flows and practices can achieve more than assessing the technical sustainability performance of fisheries and aquaculture*. A technical perspective provides relevant insights into the application of technologies and leads to prescriptions of what should be done in theory, but ignores the everyday reality of fishing and aquaculture production, processing and trade. This is not to say that technical methodologies are unimportant, but technical assessment provides only the status of a production and/or regulatory system; the underlying reasons for inaction by farmers, fishers, traders, retailers and regulators are not identified, addressed and changed. By understanding sustainability initiatives as social enterprises which seek to change the relations and practices of people rather than shaping fish stocks and environments, many of the ongoing challenges faced by the movement can be overcome.

Second, *producers are motivated by not only economic gain but also socially mediated values, knowledge and goals that shape the conditions under which sustainability is practised*. A wider set of social constraints and motivations needs to be included in the design of sustainability initiatives. Greater recognition of otherwise marginal fishing communities, the maintenance of fishing lifestyles, greater equity in rights and market values are all motivations that enrol producers and consumers to support these initiatives. However, these social constraints are not commonly used to motivate participation in environmentally focused initiatives (e.g. ecocertification) – most of which are based firmly on economic incentives and technical assessments. Importantly, this proposition is *not* a call to codify these social values, knowledge and goals into auditable standards, as most initiatives engaging 'social issues' tend to do. Instead we argue that recognising and incorporating socially mediated values, knowledge and goals into any sustainability intervention will increase the likelihood of success in shaping on-the-ground practices towards sustainability.

Third, *those who control information and knowledge determine the success of the seafood movement*. The long-term effectiveness and legitimacy of the seafood movement are already dependent on the collection, storage and communication of information. But as demands for hyper-transparency or 'extreme disclosure' (as outlined above) expand, so too will the informational demands placed on producers and other actors in the value chain. The result of these information demands will be the emergence of new actors (NGOs or otherwise) that will develop information systems based on a range of remote-sensing and mobile technologies, collect data and hold the seafood industry to account. As noted above, this is already happening in the fields of traceability and remote-sensing-based fisheries' surveillance. The implication is control over the means by which social relations and practices in the seafood movement are formed and steered. Should this control be developed in the absence of input from users and practitioners, there is a clear risk of detachment and resistance. But if users are included in the co-design of these systems, new possibilities can emerge for building mutual trust to achieve sustainability goals.

Fourth, *'sustainable seafood' will require expanding beyond fisheries and aquaculture alone.* Seafood production is only one of a huge variety of activities that occur in inland, coastal and marine waters. In the future the space and resources required to fish or farm sustainably will become more and more scarce. This means that competing claims on aquatic resources, including environmental protection, tourism, economic development, shipping and mineral extraction, will have to be addressed in order to fulfil both productive and sustainability goals. In response, we argue, sustainable seafood initiatives will have to expand their scope beyond seafood production. This poses both an opportunity and a threat to the movement. The added scope can bring further relevance and leverage by NGOs and the private sector to engage governments in a more integrated vision of the sustainability of aquatic and marine ecosystems. However, they also require new knowledge and expertise. Instead of internalising such knowledge in existing sustainability initiatives, it may be more efficient to expand the 'interoperability' between seafood and wider environmental initiatives.

Fifth, *supply-chain actors should promote sustainability beyond the supply chain.* Achieving sustainability requires more than focusing on the practices of those in the value chain. Following the previous proposition, new environmental issues emerging outside the seafood sector will become increasingly important, as will global challenges such as climate change, securing livelihoods of fishing communities and promoting processing in (poorer) producer countries. The role of large firms in the global seafood industry will become increasingly more important to address the ongoing sustainability of not only seafood but food in general. But this will require them increasingly to 'act' beyond their supply chains to promote sustainability. Such extended responsibility is already under way. But to promote this engagement the seafood movement must also adapt, by identifying and steering these companies based on a broader set of incentives than supply and consumer demand alone. New index- or standard-based initiatives that assess the broader sustainability practices of companies are one potential option (e.g. the Environmental Performance Index, the World Bank Governance Index and the Ocean Health Index, to name just a few), as are addressing the values of investors in publicly listed seafood companies.

Sixth, *private standards, codes and recommendation schemes will remain an essential part of sustainable global seafood governance.* Private standards and certification schemes have significantly contributed to (awareness about the importance of) sustainable seafood provision. These private initiatives do not replace formal government arrangements, but take up the dual roles of pressurising governments and supply-chain actors to take more effective action and filling the gaps left by national states in the context of globalising seafood provision. To enhance their impact, private standards need to be more transparent, monitor their results more effectively and be more accessible for small enterprises. The presence of multiple private initiatives may seem ineffective, but it might also contribute to increasing their overall quality and effectiveness. Private standards should, however, acknowledge their supporting position in seafood governance, and thus continue to fine-tune their contribution to the improvement of public governance.

Seventh, *states will ultimately determine the success or failure of the sustainable seafood movement.* The premise that states (especially in the global South) are not able to govern seafood production effectively remains a central driver of the sustainable seafood movement. But as we show in this book, this does not mean that states have been superseded by NGOs and the market. In fact, the opposite is likely to be true. The globalisation of food systems has pressured states to adapt to the acceleration of material, trade, knowledge and capital flows – just as seen in the seafood sector. But instead of 'retreating', states will remain essential in their role in facilitating these flows between sites of both production and consumption. Seen as such, states will continue to be the 'midwives' of the global seafood movement (Sassen, 2006), setting the conditions by which NGOs and the industry interact across global space, and also determining the types of response needed to move the sector towards sustainability.

Final word – towards (sea)food system governance

The seafood movement is one of the most dynamic food-related social movements in the world today. Few other industries have been placed under as much pressure to make changes to its fundamental production practices. Few have been given the levels of investment and attention by government and private sector alike to advocate and make these changes. This innovation and diversity are a function of the complex nature of the sustainability issues facing fisheries and aquaculture production in the world's coasts and oceans today. But to be effective, the movement will have to continue to innovate in a way that creates not only new product categories for consumers but also real change in the water.

Successful innovation will require a comprehensive approach to sustainable seafood governance that brings in different actors at different locations along the (global) seafood chain. The propositions outlined above are designed to help think through what these areas of future development will be. But overarching these is the need to balance better cooperation and competition between the constituent initiatives that comprise the movement. To be successful, they need to contribute to a collective seafood governance toolbox that can be applied across scales, countries and diverse sets of producers, traders and consumers. This will enable a more comprehensive approach to sustainability that can engage the diversity of the industry rather than reduce the industry to unrelated methodologies with different indicators and principles.

By understanding the seafood movement in these terms, we can start to understand that creative proliferation within the movement is not a problem, but is instead part of the continuing dynamism of the movement itself.

Notes

1 For more information on the Sustainable Seafood Symposium see http://sustainableseafoodnow.com/en/ (accessed 17 November 2018).

2 At the time of writing this debate is being shaped by scholars including Michael Tlusty, Megan Bailey and Peter Tyedmers.

References

Béné, C., M. Barange, R. Subasinghe, P. Pinstrup-Andersen, G. Merino, G.-I. Hemre and M. Williams (2015). 'Feeding 9 billion by 2050 – Putting fish back on the menu', *Food Security* 7(2), 261–274.

Fabinyi, M. (2016). 'Sustainable seafood consumption in China', *Marine Policy* 74, 85–87.

Fabinyi, M. and N. Liu (2014). 'Seafood banquets in Beijing: Consumer perspectives and implications for environmental sustainability', *Conservation and Society* 12(2), 218–228.

Gulbrandsen, L. H. (2014). 'Dynamic governance interactions: Evolutionary effects of state responses to non-state certification programs', *Regulation & Governance* 8(1), 74–92.

Kobayashi, M., S. Msangi, M. Batka, S. Vannuccini, M. M. Dey and J. L. Anderson (2015). 'Fish to 2030: The role and opportunity for aquaculture', *Aquaculture Economics & Management* 19(3), 282–300.

Lewis, S. G. and M. Boyle (2017). 'The expanding role of traceability in seafood: Tools and key initiatives', *Journal of Food Science* 82(S1), A13–A21.

Overdevest, C. and J. Zeitlin (2014). 'Assembling an experimentalist regime: Transnational governance interactions in the forest sector', *Regulation & Governance* 8(1), 22–48.

Pattberg, P. H. (2007). *Private Institutions and Global Governance: The New Politics of Environmental Sustainability*. Cheltenham: Edward Elgar.

Pramod, G., K. Nakamura, T. J. Pitcher and L. Delagran (2014). 'Estimates of illegal and unreported fish in seafood imports to the USA', *Marine Policy* 48, 102–113.

Roheim, C. A., S. R. Bush, F. Asche, J. N. Sanchirico and H. Uchida (2018). 'Evolution and future of the sustainable seafood market', *Nature Sustainability* 1(8), 392–398.

Sassen, S. (2006). *Territory, Authority, Rights: From Medieval to Global Assemblages*. Princeton, NJ and Oxford: Princeton University Press.

Schiller, L., M. Bailey, J. Jacquet and E. Sala (2018). 'High seas fisheries play a negligible role in addressing global food security', *Science Advances* 4(8), eaat8351, DOI 10.1126/sciadv.aat8351.

Toonen, H. M. and S. R. Bush (2018). 'The digital frontiers of fisheries governance: Fish attraction devices, drones and satellites', *Journal of Environmental Policy & Planning*, DOI 10.1080/1523908X.2018.1461084.

Index

Page numbers in **bold** denote tables, those in *italics* denote figures.

For Product Safety Concerns and Information please contact our EU representative GPSR@taylorandfrancis.com
Taylor & Francis Verlag GmbH, Kaufingerstraße 24, 80331 München, Germany

www.ingramcontent.com/pod-product-compliance
Ingram Content Group UK Ltd.
Pitfield, Milton Keynes, MK11 3LW, UK
UKHW011010180926
475701UK00004B/655

* 9 7 8 0 3 6 7 6 2 8 5 4 3 *